What Others Are Saying

Just spent a few days reading your book and found it fascinating and informative. This makes life a wondrous journey—no, an adventure of discovery. Thought provoking and evocative, this book draws the reader deep into their theological past, reminding them of their roots. It is studded with forgotten gems of religious knowledge, fanciful quips, and awe-inspiring prose.

—Patricia Graham -Award-winning poet and story writer, formerly a stone's throw from both the Irish Sea and the Lake District of Cumbria, and currently gracing the southern mountains of Spain.

What a work you have done here, Mark! I am especially attracted to the Hyksos part.

—Brad Scott , The Wild Branch Ministry (wildbranch.org). Brad is an expert in Paleo Hebrew and other ancient languages, as well as having an wonderful understanding of Hebrew thinking versus Greek (Western) logic.

Given the opportunity to reflect on the Hebrew Scriptures from any perspective is an opportunity worth grabbing. Mark Robinson's unique approach to addressing this subject manner (seeing the Almighty's hand in our lives and investigation of truths associated with the Biblical Canon unknown to most believers) is a refreshing change from the traditional and is a trip worth taking. To try and tackle virtually any opinion-based literary work on this subject is risky at best considering the diverse spiritual climate this world has experienced from the beginning. After much study and rumination in the Torah and the prophets, we have determined that the absolute, whole truth that exists in this universe is just out of reach of even the most scholarly. Nevertheless, Biblical truths seem more easily grasped when we reach out in faith. The ability to consider the world and the human experience through the eyes of our brethren is critical in understanding it ourselves. The rich, experiential knowledge that comes from living on this planet is contained in the eyes and minds and hearts of believers. Clearly, we can learn more about our Creator walking in the shoes of His other children. Mark Robinson does an exceptional job of inviting the reader to walk in his shoes…and *what* a beautiful walk it was. Well done!

—Ryan and Jennifer Farney,
Hebrew Roots Believers and Master Goatherds

While Mark was a scientist and engineer at Los Alamos National Laboratory, God saw fit to bless me, as he brought me there, allowed me to meet Mark, and then to become friends. I am a chemical engineer by training, and filled with equations and stuff that had become a frame for viewing my Master and Father and his creation. I learned about Mark a little at a time, and although there was professional common ground, his faith and his paradigms produced the depth of our friendship. As I began to read "Walking Among the Wheat & Tares...Seeking the Hebrew Roots of Our

Faith & the Hand of the Almighty [Proverbs 25:2]" his messages were so clear. It was as if Mark spoke every word to me personally. Mark explains the four ways of looking at life that he uses in terms of life lessons that allow us to see the Hand of our Creator in events in our lives. Mark is not the first to discuss the intervention of the Almighty in the course of human events, but he is one of the few in our generation to recognize his hand from many perspectives. The images he develops produce such sweetness in my heart and leave me wanting more. His simplicity and humility bring depth to my perceptions that I never expected. Perhaps I have too much of a Greek perception, and Mark sees through eyes that understand Hebrew ways. Mark has a vision of the Hand of our Creator in all the events of our lives and the lives of those who came before us. It is part of the fabric he describes, lives by, and shares with this reader. He speaks of the *B'rit Chadasha*—an appellation for the [New] covenant that warms the heart. *Chadash (שָׁדָח) or Qodesh* [*Holy*] in Hebrew can mean *new*, as well as *to renew*. Perhaps it is somewhere in between, or maybe it is both! Jeremiah 31:30 is the source. And Mark paints horizon-bending pictures with few and simple words. "If I had faith in God I could move mountains, but I don't think I'm smart enough to know where to put them." And I am so anxious to walk on through these pages that are filled to the brim with such ways of seeing, and to gather them up—at least the few I can carry—so they can reside with me also.

Traveling through these pages reminds me of a road trip—A whimsical road trip—each turn, each stop to focus, brings such fine surprises, such pictures, such poetry! The words and the pictures are woven together with hardly an interface between them! And finally, on A Closer Look at the TANAKH (the Torah, the Prophets, and the writings) waits images of the depths of our need for truth, and the need, with God's will, becomes a hunger for understanding. This treatise at the end of this tome is the stuff spoken of in the Book of Hebrews—it is the meat that belongs in our spiritual diet so we can come to grips with the reality of how Jesus/ Yeshua actually lived, the basis for his faith (the Words of His Father), without man's explanations of what is really important, and what is really outdated. Thank you, Mark, for the time, for the prayers, for the special ways of seeing things, for your insatiable hunger for the *real stuff* God has lovingly provided for us. Thank you for this book that is founded on unshakeable trust and faith in our Loving Creator.

Finally, the examinations of myths that have mischaracterized central tenets of the Christian religion and led believers into deep misunderstandings of both history and the Word of God provide great illumination that has been hidden for centuries and are very, very, welcome back in the light.

—Don Mullins,
Town Crier and Lamplighter for
Northern New Mexico

I have known—and not known—Mark Aaron Robinson for forty-five years. Unfortunately, he vanished into the ether right after high school. This year our classmates asked me to search him out so we could invite him to our reunion. As a consequence of my successful search, I am honored to write this review of his insightful, and, if I may, *whimsical* manuscript. I find Mark's poetry profound in it's portrayal of the intersection of God and man from the beginning to the end, which he describes in the *Vox Humana* section as *In the interim between dust.* More than that, the poetry also speaks of the good in life, warms the soul, fills the heart with whimsy and laughter (In *retrospective* made me laugh out loud), and flavors tears with salt that we can't live without. And the illustrations take my breath, while they fill up my eyes, and tell no lies. In the B'rit Chahashah discussion, Mark writes, "He has written us wings, and, unburdened by things, we can fly." I would have said—Not I!—before I read this manuscript. Now I'm a believer. God bless free riverbanks and God bless Mark Aaron Robinson!

—MJ Kern, Follower of the footsteps of Johnny Appleseed & Finder of Lost Persons

Walking Among the Wheat & Tares

Walking Among the Wheat & Tares

Seeking the Hebrew Roots of Our Faith & the Hand of the Almighty [Proverbs 25:2]

MARK AARON ROBINSON

TATE PUBLISHING
AND ENTERPRISES, LLC

Published by Tate Publishing & Enterprises, LLC
127 E. Trade Center Terrace | Mustang, Oklahoma 73064 USA
1.888.361.9473 | www.tatepublishing.com

Tate Publishing is committed to excellence in the publishing industry. The company reflects the philosophy established by the founders, based on Psalm 68:11,
"The Lord gave the word and great was the company of those who published it."

Cover design by Junriel Boquecosa
Interior design by Jomel Pepito

Published in the United States of America

ISBN: 978-1-63122-795-0
Religion / Biblical Studies / General
15.03.23

Dedication(s)

To Ronald Farney, husband, father, teacher, artist, brother, and brother-in-law with a heart so big and bold.... He never got old

To our good friend, Timothy Daniel Mouyos. Dan had a soft spot in his heart for every mortal, but held arrogance and hypocrisy in pure contempt. If anyone ever really knew that the shortest distance between two points is a straight line, it was Dan, but he would go miles out of his way for a friend. He was surely one of the best of us, and he is sorely missed by many.

1968, Helen Fitting, Hiram and Wellesley, Master of Arts,
"Dear Rocky, I shall keep the memory of
your multi-faceted personality and character
always, not to mention your eyes
that reveal a thousand things." RIP 1987

Table of Contents

About the Author

Headwaters of the Jordan, 2011

Mark Aaron Robinson is a scientist, engineer, and author. He has traveled the world (Austria, Switzerland, the Philippines, the Hashemite Kingdom of Jordan, Kenya, and Saudi Arabia) solving problems with radioactive and hazardous waste for the International Atomic Energy Agency. His first book, *100 Grams of Uranium Equals 290*

Tons of Coal, ISBN 0-945318-00-6, 1987, sold in thirteen countries and was a staple in libraries of the United States and Canada. He has also been published in a number of IAEA Technical Documents, Nuclear News, The Journal of Nuclear Material Management, and Popular Science. His literary works also appear in four books published by the International Library of Poetry. He has been happily married for more than forty-three years and is the father of two children and the grandfather of three.

Preface

This book is comprised of twelve parts:

1. Considerable verse that arises through inspiration from wonderful literature teachers and professors I was blessed by, and inspiration as a function of walking on the land where another walked on the water. All of this, I see as a manifestation of faith in my Redeemer and his Word (mostly misunderstood by many). The subject matter is my life experience from 1970s through today. Much of my inspiration comes from living on the Mississippi River, canoeing the boundary waters, the French, and Wanapatai Rivers in Western Ontario.

2. My occupational experience at nuclear plants (Perry and Trojan), the plutonium facility at the Los Alamos National Laboratory, the Waste Isolation Pilot Plant outside Carlsbad in New Mexico, and investigation of issues with radioactive materials on

several continents at the behest of the International Atomic Energy Agency in Vienna, Austria, that has broadened and influenced my world views.

3. I solved problems with radioactive and hazardous materials in the Philippines, the Hashemite Kingdom of Jordan, Kenya, Austria, France, and Saudi Arabia. This became possible as part of my functions were solving transuranic, radioactive, and hazardous wastes issues in the United States while serving as a Professor in Hazardous Materials Technology. My accomplishments in these venues were possible through my education in Ecological Engineering from the Colorado School of Mines, Master of Science Program, and my doctoral work in Environmental Science at CSM and Lacrosse University, and hazardous and radioactive/nuclear materials law studies at the Denver University School of Law. Through my education and experience I became valuable to the International Atomic Energy Agency when it comes to evaluation and solution of radioactive materials and equipment issues in Third World countries on several continents.

4. An in-depth look at the TANAKH (the Old Testament) showing how to find the Messiah [Ya'Shua, which is Hebrew for God's *Salvation*] or Jesus as he is known in the Christian community,

therein, and a discussion of the origin of paganism in Christianity.

5. An examination of the extant record, the Hebrew scriptures, and literature produced by Egyptologists to determine who the Hyksos were and what was their derivation.

6. A demonstration that ancient records of Egypt confirm the Exodus story.

7. An examination of the age of the Earth via historical events and radionuclide decay measurements.

8. An examination of the Hebrew names of the Messiah and God the Father

9. A determination of the birthday of the Messiah (which is certainly not December 25).

10. An examination of pagan gods and goddesses and their influences on Christianity.

11. A look at the function of the Fishers in scripture.

12. An examination of the influence of pagan gods & pagan practices on Christianity.

A GLANCE AHEAD

- A painting by Carla Sue Robinson is the source of the illustration for *Eutrophos.*

- Zoe Anna Robinson is the creator of the dragon illustrating *Mystery/Replacements.*

- Illustrations for *Retreat (Moss Haven), My Own Volition, and Foregone Conclusions* are used by permission of the National Geographic Society. They were extracted and/or modified from photographs originally published in *Exploring America's Backcountry*, 1979.

- Mark Aaron Robinson created the balance of the illustrations. Many components of these illustrations were extracted from *Deskgallery Megabundle* by Zedcor, Inc., Tucson, Arizona.

- Rodney Farney, Jr., well-known western artist of Salida, Colorado, created the illustrations for *Metariverism, Salmon Run, Earthbound/Skyblue Bondage, Snow,* and *Good Question.*

- Personal Note: I'd like to point out that whatever literary and science skills I may have, I believe I owe to the good works of my ancestors. I am descended from Pastor John Robinson whose five sons came to America on the Mayflower in 1620. See *Robinson Lineage From Parents of Reverend John Robinson*, daughters of the American Revolution Lineage Book Vol. 137, page 31; Vol. 33, page 99, and Vol. 128, Page 218.

Foreword

I've ripped up your book into little pieces.

Some of them I've put on the wall as reminders of what's important in life. On three pages, I've bent the corners. Some of them I've folded in half as a reminder to write a song using your poetry as lyrics, though quite probably I'll never seem equal to the task.

You've asked me to write a *frontispiece* (yeah, perhaps better called a "frontispeace" given your deep relationship with the always-available-but-seldom-invited Grace) and I assume that the matter is herein accomplished...

but look at me:

> I'm so caught up in your way of writing/thinking
> that I find myself stealing the form of your thought
> (certainly not the substance)
> and hoping you'll remember that
> imitation is the sincerest form of flattery
> and for you to forgive...

Now I'm circling your paragraphs with black ink and drawing arrows to others...I'm surrounding the flow of your thought with graffiti like *wow* and *amen* and *absolutely* and I can't help myself though I know that these comments would not be welcome to my own eyes were I reading it again for the first time…

How dare someone
forecast the effect on my heart!
When I read a foreword,
I don't want to know my reaction beforehand,
but only if the piece is done well.
Still, it's too late for these pages;
passion has sung her song
and the cost is an echo of enthusiastic ink
along the borders.

I won't be lending out this copy...
my simple advice to your readers:

Celebrate your own!
By Noel Paul Stookey

Noel Paul Stookey is, of course, Paul of *Peter, Paul, and Mary*, who blessed the world for more than fifty years with sixty folk song hits. Paul's classic work, "The Wedding Song (There is Love)", has helped literally millions of lovers encounter the Author of Life in their wedding celebration. Noel is also the founder of Hugworks (originally

Celebration Shop), a nonprofit organization devoted to the musical encouragement and support of children dealing with severe, chronic, and life-threatening illnesses. Noel, Celebration Shop President Jim Newton, Paul Hill, and others combined their talents to produce *Friends of the Family*, music for children, which won a Parents' Choice Award in 1990. Noel and his wife currently produce multi-faith seminars and performances. I am very thankful for Mr. Stookey's kind input and the lasting good works of Noel and his wife. And, I still get misty–eyed whenever I hear *If I had my Way, Michael Row Your Boat Ashore,* or *Don't think twice.*

Mark Aaron Robinson

Noel Paul Stookey

Introduction

> When I was a child, I spoke as a child,
> I understood as a child,
> I thought as a child: but when I became a man,
> I put away childish things.
> For now we see through a glass, darkly,
> but then face to face:
> now I now in part, but then I shall know
> even as also I am known.
>
> 1 Corinthians 13:11-12

Who among us can fully understand our journey, at once carnal and yet divine, with our individual destinies so shrouded in mystery, so influenced by intangibles that a tentmaker described so profoundly? Clearly, we cannot see clearly—ourselves, our world, our beginning, our end, or our Maker. But we can see in part. And it is this imperfect vision of the infinite that I have been struggling all my life to describe.

I spent a lifetime on the banks of many rivers that are in the global scheme of things, just one. As I am by education an Ecologist and an Environmental Scientist and because I have lived and worked on the shores of many rivers and lakes in the United States, in Canada, in Israel, and in the Kingdom of Jordan; I have in the past referred to these streams collectively as the "Metariver Inn." Many a fellow traveler I met on the lee shores—lives intersecting in four-dimensional paths from the past and into the future. Fellow sojourners in an inn without stasis to who we shared anecdotes and epiphanies around the kitchen table of life and departed to unknown destinations. This then is the resulting collection of morning coffee observations and evening ale inspirations of travelers diligently seeking or desperately circumventing the real threshold.

Of this I am confident: the time when there was no room at the inn is a *fait accompli*. Whosoever will, may enter now. But description of the inn is best left to the Irish who have valued poets above kings since the advent of language. So like a blind Irishman sorting potatoes, I have relegated my tales to five chapters—or, if you will, five ways of looking at the universe—based primarily on my sense of their probable digestion as I held them in my hands...*Caed mil faelte*!

Metariverism is a study of our reflection in "the river" and other bodies of water (in the ultimate biological sense we are all "bodies of water"). It is spiritual limnology. It is the widely held belief of one man that a river is not merely water, but a flow of analogies. It is the sum of years on land's edge. It is my metaphysical view that, although water is merely molecular obedience to laws of the universe that we have just begun to find out, matter and constants have a design that reveal many truths. I am a seeker of such sooth, and my verse, hopefully, is all wet —and thus alive, for everything desiccated dies. Like a river, I cannot make a beginning; my words fall from my memories and consciousness to become a part of the flow of my life. And my truth is like a handful of water—I cannot hold it unless I'm in the river, and then it's really the river that holds me. When I'm waist-deep, I'm half-full. When I'm full, I'm over my head. There are only moments when I can merge with truth till the mortal in me gasps for familiar terrain. If God made both the river and I, then there is much to be seen by looking into both of us. The difference between us is that I ebb and flow, through changes I go. I see clearly and dearly then I am weary and dreary. But the river doesn't shiver and it doesn't sweat—it's just always wet. Yet it lives and gives and laughs and rafts and cries and dries and just passes by—like time, like rhyme, like a life like mine. We are both an image on God's eye, only he doesn't blink and he doesn't lie. To see through that Lens—just in tacit facets and through mortal portals—is all that I'm trying while I'm living toward dying.

Anti-riparianism

The riparian environment is that expanse of land adjacent to the shoreline that is influenced by a body of water. The riparian environment has very obvious boundaries in an arid land. In fact, in a desert or chaparral, the watershed is interrupted—any moisture penetrating the ground is wicked back by the thirsty surface, bringing also a baggage of salts and alkali and that which are anathema to life. Conversely, in *rainfluenced* environments, riparian edges are less distinct and more subtle. What is certain is that life close to the source of life is the most abundant, and that mankind has a damnable desire to shut the valve, to ration the flow. In the business of getting and spending we use all means at our disposal to negate the riparian influence and to create our own confinement. Such tendencies are consistently anthropomorphic and resultant destinies are insistently less than our Maker intended. I call this carnal characteristic *Anti-riparianism.*

B'rit Chadashah

B'rit Chadashah is the Hebrew term for the New Testament. In my view the disciples and the remarkably many believers demonstrated faith and courage beyond anything we have seen in history. I have tried to portray such courage and faith in the verse that represents this book of faith, courage, and love…love of the Messiah, love of God's Word, love of disciples, and love of the Torah (rules for living given to all believers by the Almighty through Moses). This section of the book focuses on the mighty men of faith whose deeds are recorded therein.

Footsteps of the Prodigal

As you might guess, the verse in this section flows directly from my own life. What is not so readily apparent is how common, how recognizable, these experiences are to my fellow travelers. The truth is that we all learn the same lessons, or die trying. Saint Paul (Rabbi Shaul) said there is no temptation that has taken you such is not common to man. Solomon said there is nothing new under the sun. Santayana said that he who will not learn from the past is condemned to relive it. The point here is that we are all prodigals to some degree from the moment we first set a course for independence. So I like to think of this verse as a wave from a stranger on a bus bound the other way. I know I'm a lot like him or her— we all have the same beginning and hope for the same end.

Generations

Somehow my wife and I survived the insanity of the 60s, and followed God's plan for man…We got married, had kids, worked hard, and became respectable. I worked during the day and went back to several colleges at night or whenever

I could fit it in. I finished a bachelor's in engineering, later a Masters of Science in Ecology at Colorado School of Mines, and a decade later I went through the Colorado School of Mines PhD Program in Environmental Science (but ran out of leave from my employer and had to go back to work). Finally I got a PhD in Environmental Science at Lacrosse University. I was then assigned to a multitude of opprobrious tasks that were waiting for someone who was considered deserving, given the blessing of time off for post-graduate work. About the same time our children were struggling through college, or jumping straight into the technical work force, and getting married themselves. One generation follows another, and that is what this section of my tale examines. Our conclusions (my wife and mine) can be boiled down to this: it's tough to raise kids and get them through school and into the workforce, but all is forgotten and forgiven when grandchildren show up!

Non Illegitimi Carborundum

In the vernacular, this translates as: Don't let the bastards get you down! This should be taken as a command from a father and good advice for anyone...here in the *sixty-first century*. I use it here as a topic appellation befitting the evildoer, the city sewer, horsemanure, plastered bastards, the worst of the unhearsed, and all that's unpleasant, plebian, and aeolian.

Vox Humana

I remember a four-hour essay exam during which I got a nosebleed and left great drops of life all over the test papers. My professor thought it was evidence of sophomoric self-flagellation and wondered emotively in her grading regarding my reasons. Though her assumptions were wrong, she was intrinsically right, considering the unbreakable link between the conscious, unconscious, and human physiology, We cannot shriek without bleeding and we cannot bleed without shrieking, whether air is modulated or not. In the interim between dust we experience the "human condition", and no one has yet paid enough dues to sing those blues definitively. From this pen flows the ink of a thousand generations encountering the quandaries of existence. I am more than the former gleam in my father's eye—I am the sum of all their expectations....and yet I am alone, as alone as a man can be in the midst of a universe of inanimate atoms and where God is beyond matter. In the face of such unfathomed isolation I can only gesture in the hope of communicating. It's a brave smile and a silent sob I call *Vox Humana*.

Metariverism

and a river went out of Eden to water the garden..

Revelation 22:1

Mine is the sunlight, mine is the morning, borne of the one light Eden saw first.

Morning Has Broken (hymn)

And he shewed me a pure river of water of life...

Genesis 2:10

Jordan River is chilly and cold, chills the body but not the soul

Old Spiritual

Let us pass over to the other side.

Mark 4:35

So how did Huck merge with the river?
He slept on it, he swam in it, he dwelt on it!

—Helen Fitting, Hiram & Wellsey, M.A. English Literature, R.I.P Ca.1985

Old Man River, he don't say nothin',
He don't plant taters, he don't plant cotton,
and them that plants 'em is soon forgotten.
But Old Man River, he just keep rollin' along.
—Jerome Kern, Showboat

We were neck deep in the Big Muddy
and the big fool said to push on...
—Folk Song

And so we beat on, boats against the tide,
borne ceaseless back into the past
—The Great Gatsby, F. Scott Fitzgerald

A man may fish the worm that hath eat a king,
and eat of the fish that hath fed of the worm.
—Hamlet, W. Shakespeare

I heard the old, old men say,
All that's beautiful drifts away, like the waters.
—W. B. Yeats

All good writing is swimming underwater
and holding your breath.
—Undated Letter, F.Scott Fitzgerald

Old men admiring themselves in the Water
—W. B. Yeats

Metariverism

With the will of a waterwish
and the gill of a genufish,
in the flood of the feverish,
I'm over bankful,
and thankful.

What does the river mean to the shores it flows between? A fisherman said, "Fish eat worms and worms eat dirt, and dirt eats everything." So I say the river serves the earth a long drink of water, a chaser, an eraser, a very warm glacier.

But what can the river mean to the scores who glean its gleams?

A wisherman says, "I'm a dreamer, a sit-beside-a-streamer, and what I sit by seems my Redeemer: It matches the sky, declaring the glory of God, and enters my eye, where many have trod and untrod."

When I take a mouthful I wind a whirlpool around my similes and swallow like the earth in motion. When I take a mindful I find a pure fool piering in river darkly, waiting for the Perfect to arise and blind my eyes like dawn upon the water.

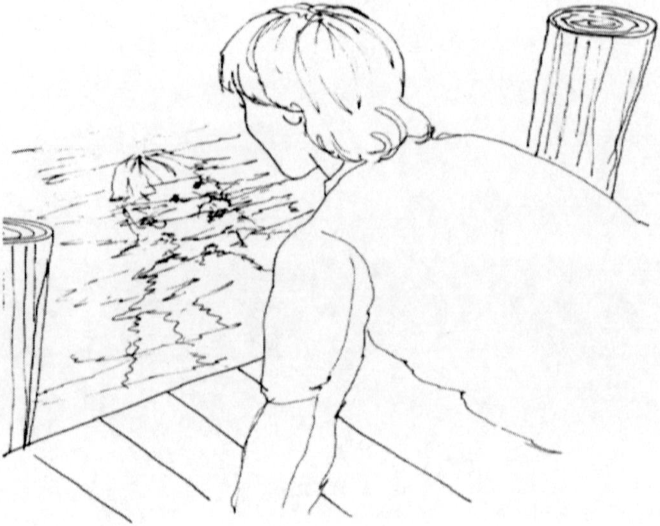

Yes, I teeter, I totter, mostly I otter

-living in the river and on the land, looking at God through the eyes of a man.

I'm a note in a bottle I can't read myself, afloat on a mirror of infinite stealth, that ripples truth with myriad waves past men with computers who came out of caves, learning but little in daze after daze.

The river runs into the sea, yet the sea is not filled. Time flows too, and cannot be stilled. Truth is sheathed by time and tide, yet closely wreathed with the riverside. Beauty covers it like the sky upon the water. Pain hides it in a lamb led to the slaughter. Life defies it like today belays eternity. But the river defines it certainly: the Promised Land is on the other side. God it's wide.

Peter, sinking, reaching for Jesus, and Jesus Walking on Water, *Ivan Aivazovsky Painting*

The Rock on Water

How I love my rowboat...
my three-point landings on the waves,
the wind upon my old coat
is the wearing that it craves,
wind that fearsome blows me
far from pier and now
to waters where I'm told he
got out by the bow somehow.

As my aching oar strokes
his footprints on the water,
only a watershed cloaks
his steps and how they falter.
I see his hesitation
as he looks down at his feet.
He sank in desperation
and that glues me to my seat.
And yet his feeble hand out
was held, through sink and swim
—it helps me row my heart out,
with a grin, against the wind.

—M.A. Robinson
Originally published in
Our Twentieth Century's Greatest Poems,
World of Poetry Press (1982)

Is There Frog Heaven?

Fred Frog left his bog,
a pedigreed pollywog.
What are the odds
he'd become like the gods
—knowing good from evil,
but needing a retrieval?
Needless to say,
he went the frog way,
and leaping of course
into space-time warps,
he goggled all history
—how is a mystery.

Now history in part
will make a frog fart,
but history in total
unveils the anecdotal.
Fred could only stare
beneath his fine frog hair!
With a croak and a quake
he jumped back to the lake,
and the frog who saw all
found himself at the Fall,
in the mouth of his advisor
who we know still lies here.

Slowly, inexorably,
it swallowed Fred horribly.
Cold and relentless,
it hates what it ate,
but is grossly repentless
of unwilling weight.
And there is no salvation
on this amphibious occasion.
If I were a frog
I'd stay in my bog
and remain in denial
to every reptile.

Eutrophos

The banks of a river
are the legs of a woman,
and she has known
her share of yeomen.
Some swim in the river
and she receives them.
Others insist she deliver
whatever they believe in.
Upon these banks
arrive extremes then.
Only the hydrophilous
thrive between them.

Of their beginning,
men have theories and nomenclature,
all too dreary and detrital in nature,
while the river flows from her inception,
and Heaven knows by what direction.
Between those banks I understand
only a fathom of what is planned.

Anti-riparianism

To dig and delve in nice clean dirt can do a mortal little hurt

—Gardening, John Kendrick Bangs

The black earth drinks and trees suck rain. The sea drinks brooks and the sun the sea and the moon the sun.

—Anonymous, (Anacroen, Odes No. 21)

Any nose, may ravage with impunity a rose.

—Sordello, R. Browning

It's a comfortable feeling to know that you stand on your own ground. Land is the only thing that can't fly away.

—The Last Chronicle of Barset, Anthony Trollope

No man can lose what he never had.

—Izaak Walton, The Compleat Angler

He is useless on top of the ground. He ought to be under it, inspiring the cabbages.

—Mark Twain, Pudd'nhead Wilson's Calendar

Pity the only way to paradise is through a hearse.

—Stanislaw J. Lec, Unkempt Thoughts

The end of the human race will be that it will eventually die of civilization.

—Ralph Waldo Emerson

...You take paradise, put up a parking lot...

—Joni Mitchell, Big Yellow Taxi

We build machines that we can't control
and bury the waste in a great big hole

—Sting

You have far too much ego in your cosmos.

—R. Kipling, Life's Handicap: Bertran and Bimi,

He that is giddy thinks the world turns round.

—W. Shakespeare, The Taming of the Shrew

It's a poor memory that only works backwards.

—Lewis Carroll

Quick, Throw me some existential meaning!!

I lost everything

I went for a swim,
I just wore my skin.
I left my shirt,
I left my wallet,
I went right in
up to my what-ya-ma-callit.

The river got taller.
I got smaller.
Suddenly I was gone.

Certainties

Winter is when I row
—the Mississippi in December
is white caps and unsteerable winds.
My oars tread the waves
with strokes of joy.
Shores disappear under current possibilities:
the Lord could calm the river,
the Lord could call the liver,
I could live to row another day.
Hooray.

Fishing at Redbud Isles, Eve Larson gallery

Little Whiles

Dawn dissolves as morning flows
down daylong dreams of Heaven knows
what wild streams that suddenly froze
in still life scenes of what life means:
A time to hold and then let go,
once a child and once we're old,
miracle miles on short long roads,
smiles and trials for little whiles.
Till twilight grows from amber glows
and sunset sighs in our souls.
Soon sunrise becomes sunrose,
beheld by eyes it mesmerized.

Heart River, North Dakota

Heart

O River, kind mirror,
let me be what thou art:
an amplifier
of the ample fire
of the heart.

The impossible missions are the only ones which succeed.

—Jacques Cousteau

An Ocean Remains

I wish my face
had the intensity of my heart.
I want to convey so sensitively
—yes, impart
—feelings from the depths of me.
But like the source of the sea
I am poorly perceived
by souls of the surface who've only believed in images.
Yet the depths have been pierced
in one lingering look
by those who received because they forsook their illusions.

Blah,
Blah,
WELDING
Make $5 to $15 a Day

Rowboat Therapy

(Sink right down to the bottom; rise right up to the top)

It's been a bedlam day.
Words of welders reach me
only after long delay,
though they beseech me.
Their frazzled faces don't fathom
the river where I am,
undertowed, under heaven,
going down like the damned.
Though my head is swimming
my fish ears hear
the day din dimming,
so I pull for the pier.
On my rote ride home
I'm launching my craft:
a voice beyond my own
is calling me aft.
At last I'm looking
back at the bank.
I am the Oarking
rowing out to thank
the rivermaker.
On the meniscus of a world
—what a givertaker—
I am bouyed and uphurled:

where gravity pulled me down,
I am a man caught up,
and I boat without bound
in an overflowing cup.

Even if there was once...

In the final analysis,
is the freedom of choice
or voice
as important as freedom
to go down to the river and fish?
Don't I wish!
Down with the oppressors of fisherman
and God bless free riverbanks!
Freedom to touch the earth
and give thanks
outranks
mere freedom to hear or tell of freedom.
You can touch fish and fishers too,
or throw them back if you want to,
but it's hard to grasp concepts without substance,
even if there was once...

Salmon Run

Embryonic urges
toward final fish dirges
serve salmon surges
—of strength beyond endurance,
of speed above the current's
—to return to the headwaters
like rain from the shedwaters.

The future's a conception, philosophers say,
and it's a dimension in a mathematical way.
But to salmon it's a river of days made of day.
I have no memory of being born,
and only a navel to show I was torn
from the infinitesimal embrace
of two cells in a stream,
that swam in a race
of passion and dream
that ran two directions
like all our perceptions.
And now that I've come
To a land I forgot,
I juxtapose striving with striving not.
In this earthbound kingdom I've come from the dust,
to marry and bring some life from my lust,
on my way to returning to where salmon must.

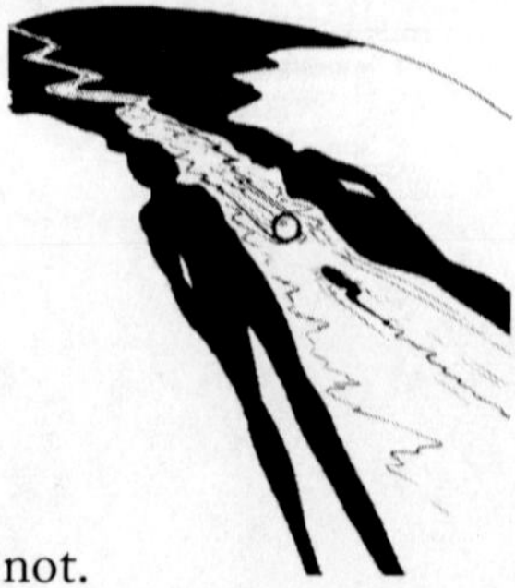

(This may really be a bridge too far over an ex-stream
theme, even though we are more than what we seem
If I live just to spawn and die,
I would be a sub-syllogistic guy.
and never really wonder why.
Yes, I was born on earth and the place of my birth
holds nostalgic attraction, but I swim in a river of time,
and the real "bottom line" is no hermeneutic abstraction:
Salmon are just future detritus, whereas I will rise
and God will baptize when it's time to invite us.

Few Cheers

Banks are bare beside the river,
where delta farms are done forever.
Gone are the groves where the waterfowl flew,
silent is the sound of the music they knew.
Erased are the fields that drank the rain,
displaced are those who grew the grain.

Concrete came and is accursed
by the earth that it enhearsed,
covering the former farmer's futures,
smothering the former farmer's few cheers.

"STOP"
NO RETURN
RESERVOIR
ROCKY MOUNTAIN WATER WORKS
MEN
MUNICIPAL TREATMENT PLANT

Good Question

I feel so far from the river
—like distilled water,
sterile and bottled up,
to flow from a paper cup,
through an executive's kidneys
and beyond...
through the manmade watershed
to the treatment plant
where flocculation will precipitate
the heavy metals
and the centrifuge will dewater the sludge.
Though I'm ultra-filtered,
will I ever feel clean enough
to again run off the highlands?

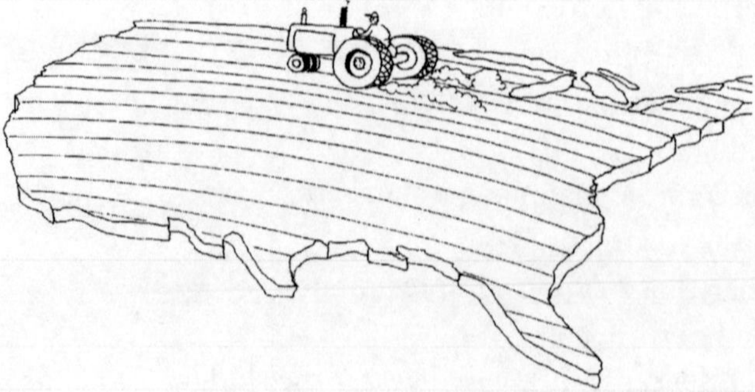

Snow

Iowa isn't everywhere,
or else they'd build tractors
for coast-to-coast plowing
and people would die of corn-boredom.

While everyone sang
of fertilizer in the Spring
I'd move south
and pray for snow.

B'rit Chadashah

Another parable put he forth unto them, saying, The kingdom of heaven is likened unto a man which sowed good seed in his field:

—Matthew 13:24

But while men slept, his enemy came and sowed tares among the wheat, and went his way.

—Matthew 13:25

But when the blade was sprung up, and brought forth fruit, then appeared the tares also.

—Matthew 13:26

So the servants of the householder came and said unto him, Sir, didst not thou sow good seed in thy field? From whence then hath it tares?

—Matthew 13:27

He said unto them, An enemy hath done this. The servants said unto him, Wilt thou then that we go and gather them up?

—Matthew 13:28

> But he said, Nay; lest while ye gather up the tares, ye root up also the wheat with them.
>
> —Matthew 13:29

> Let both grow together until the harvest: and in the time of harvest I will say to the reapers, Gather ye together first the tares, and bind them in bundles to burn them: but gather the wheat into my barn.
>
> —Matthew 13:30

> And there came a voice out of the cloud, saying, This is my beloved son: hear ye him!
>
> —Luke 9:35

How do you tell a tare from wheat when they appear identical? The tare is poisonous. It mimics the wheat in appearance. But when the wheat is ready for harvest, the weight of the kernel causes the wheat to bow. The tare is proud and does not bow because there is nothing of value in its head.

The B'rit Chadasha is filled with parables of hidden wisdom that only the wise and the good and faithful servants seek out and find. In fact, the entire Bible has hidden wisdom that only the diligent seek out. May you all become such seekers.

Ya'Shua the Messiah declares James and John to be the Sons of Thunder

Boanerges

(a colloquial interpretation)

James and John were walking along,
to or from fishing, one might expect.
They might have been singing a song,
planning, or wishing,
or with due respect,
thinking the thoughts
of men with wives
that look forward to evening
most of their lives.

Maybe the hair stood straight
on the back of their necks.
Maybe time slowed down
like during camel wrecks,
or plagues of insects.
Doubtless their lives
passed before their eyes,
all the truth and beauty,
hurt and lies,
we give too little credence
and too much disguise.
But there is no deception
in immaculate reception.
Jesus turned them aside

from the path all-to-wide,
and showed them the Christ
on His way to be sacrificed.

Transfixed in an instant,
and in time transformed,
these former fishermen
fearlessly stormed
Satan's throne
and the walls of Rome,
without a jawbone, or even cell phone,
armored only in faith
and a new concept of Home.

Like lightning before the holocaust,
and thunder from the Pentecost,
James preached to the listening lost,
pissed-off the boss agnost,
and gladly paid the mortal cost:
Herod put him to the sword
while on his way to be adored
in all his finery.
But when men called him divine,
his oratory turned to whine,
and he was eaten by worms,
from inside, before he died,
in graveyard irony.

John spread the Gospel
in a world that was hostile.
Caesar boiled John in oil
and it caused such turmoil
(because you cannot fondue
a really faithful Hebrew),
Rome banished him unto Patmos,
a place for the wholly asbestos,
where John saw farther
than silly Caesars could see...
to the end of history.
Yes, while Caesars killed subjects
(in ignorance of Hegelian dialects),
John was given Revelation,
time travel for believers
of each generation,
and deluding der fuhrers
of this Dispensation.

But, John, the second son of thunder,
wrote not what the seven thunders
thundered back then,
so I expect to see him prophesy again.

For Hebrews, to make Aliyah is to return to the land of Israel. Abraham and his people crossed over and dwelt in Haran.

The Meeting of Abraham and Melchizedek
1464-67 Sint-Pieterskerk, Leuven

Note: Melchizedek is derived from the Hebrew words Melech Zedek which means King of Righteousness.

Aliyah Bet

(the second return)

The children of Judah and Ephraim never meet,
though the long lost brothers pass in the street,
free from *Mizraim*, but not from its dream.
There may be *deja vu*, but no inkling that we two
share genes back to Noah, and certainly the *Sho'ah*,
seen from two sides, though neither presides.
Ephraim believes he is a gentile,
and Judah can be xenophobe or xenophile.
Yet sometimes, we hear our names in the breeze,
and flashes of very old memories…
prayers we didn't pray, and words we didn't say…
seem like our own, as we struggle to atone
—drawn toward a day of blessing and cursing
the world will shudder to divine,
though it's been rehearsing the cursing
and professing the blessing, a long, long time.
Whether we choose to die or to live,
we have breath now, and testimonies to give,
by word and by deed, though we tremble to read
our anthologies: life stories of every generation
ending only two ways—in sin or salvation
—as the children of two *diasporas*,
seek without seeing, reunion that was,
foretold for days we think anachronous.

*Originally published in The International Who's Who of Poetry, International Library of Poetry, 2007 For Hebrews, to make Aliyah is to return to the Land of Israel. Abraham and his people crossed over and dwelt in Haran.

Perhaps the Apocalypse

It's a seminal generation, desirous of a throne,
that doesn't really care, who or where, it kisses,
enrapt as it gazes, through the eyes of Narcissus,
on its own images, cast in pixels instead of stone,
and the foundations of its leisure are no longer known.
We have eyes, and grand designs,
but don't really see the signs....
bread, circuses, and STDs for the masses,
idols, ecstasy, and envy among upper classes,
fire, flood, and frenzy from the *madrassas*,
more and more and more morasses...
So the harangues of archaic atavists fall on deaf ears
as just hindering a destiny unachieved by their peers.
But history is punctuated by generations so blinded,
they perceived what they believed
as demiurgic and high-minded, and statues and edifices
raised to their grandeurs remain as ancient unplanned
sepulchers amid the bones of minions and emperors.
Still, stars in black cars, and an inevitable great orator,
jockey for positions in the next World Order, or,
whatever ism and shibboleths they spawn,
but none think in terms of right and wrong,
and good and evil are concepts long gone
on the road to new kingships, and conscripts,
and perhaps the Apocalypse.

Originally published in The Best Poems and Poets of 2004, International library of Poetry

Adam and Eve, Lucas Cranach the Elder, 1533
[Public domain], via Wikimedia Commons

Reading from Romans outside Laodicea

Adam said to Eve...You are bone of my bone & flesh of my flesh...a metaphor of the miraculous made of facts of the ground by the ancestor of all of us who was inspired & profound before he knew good from evil. Today, Adam has six billion living descendants who've complicated life more than a little. In limited space and time, we intersect, avoid, and miss each other, predominantly influenced by our fallen nature, & only occasionally aware of the still small voice, even though we can chose to listen. Like Adam, I am just dust, empowered by a union of synapses & spirit too wonderful for explanation. But life is too complicated for me to distill into a single sage observation. The best I can hope for is utterences too deep for words.

The Best Poems & Poets of 2002, Poetry.com

Earthbound/Sky Blue Bondage

Old Harry was his fence,
knee-deep in the ground he shouted,
"You'll never get in here!"
Harry's in deeper now.
Joe liked Harry's fence.
He thinks he's a good neighbor.
Joe doesn't build fences though, he drinks his.
Some folks build up, Some dig down.
We seem to lack a sense of elevation,
Tied to damn dirt for our foundation.
Everyone knows the longest way to your Maker
is through dust.

I've been the dumb end of a graveyard shovel.
I've seen the soon end of a hermit's hovel.
I don't want any Déjà vu…I'm not that jaded.
And my thirst for what's true has not been abated.

These silhouettes are cut with a knife
from fruitless deathettes against the Author of life. He has
written us wings, and, unburdened by things,
we can fly. But men curse their fate: that they should die
and not predicate, death with possessions.
Oh, Lord, with the eye of an Eagleman
from the feet of a feeble man,
I soar,
from the breast of a brown land,
to the bosom of Abraham.

Hopeful Calvinism

I don't want to be the architect of tomorrow,
or even responsible for today.
With so may alternatives,
but really just one,
I can't fathom the consequences
though I know there are some.
Based on my performance so far,
I think I'll refrain from selecting
any ultimate destinies,
hoping the right one will choose me.

Via Crucis
Via Lucis
Vaya con Dios

Faith and Geology

If I had faith in God
I could move mountains,
but I don't think I'm smart enough
to know where to put them.
I will wait upon the Lord
to arrange all changes.
If I had faith in God
I would not be impatient
for such metamorphoses,
yet I will wait upon the Lord,
impatiently.
Being an environmental scientist,
I know that mountains move
one atom at a time
as granite weathers into clay
and clay tries to prevent further transport.
So even though mountains move
in geologic time,
God moves mountains
by moving peoples and borders
Around them.
I'm more than ready
to be so moved,
knowing also,
that God occasionally moves mountains and lives
rather explosively.

Bible Codes

Bible Codes Link the ancient Kingdom of Judah in Israel with modern day Ireland. Also, Ancient history of Israel, and historical documents of the Irish kings, indicate that the prophet Jeremiah took Tamar, the daughter of Zedekiah, to Ireland to escape Nebuchadnezzar II , where she became the matriarch of the line of ancient Irish kings. Jeremiah is also credited with bringing the stone that Jacob used for a pillow when he was given the vision of angels climbing up and down a ladder to heaven. That stone –the Stone of Scone has been used in the coronation of Irish, Scottish, and English kings for thousands of years.

Bible Code matrix linking Ireland to the ancient Kingdom of Judah in Israel [from Yair Davidy's Brit-Am.org website]. In general in our Brit-Am Bible Codes analysis was found to be strongly associated with Judah in some way or other.

This is the lowest result received for Ulster at a reversed sequence of 521. Begins at 2-Kings 21:4 where it speaks of the deeds of King Menasseh of Judah. Crosses 2:Kings 17:20,

20:10 20:2, 19:30, 19:21. In the matrix we have emphasized the expression “REMNANT THAT IS ESCAPED OF THE HOUSE OF JUDAH” where the last letter of the word-root “REMNANT” forms part of the name ULSTER:[2-Kings 19:30] AND THE REMNANT THAT IS ESCAPED OF THE HOUSE OF JUDAH SHALL YET AGAIN TAKE ROOT DOWNWARD, AND BEAR FRUIT UPWARD.

In the matrix we have emphasized ULSTER intersected by the LAND OF CANAAN, THE COAT OF JOSEPH (of Tartan color and design?), and the name JOSEPH. In the Bible “ LAND OF CANAAN” is the usual term for “Land of Israel”. Many settlers from Scotland came to Uster and the Coat of Joseph was a Coat of Pasim meaning variegated colors in a tartan (Scottish type) design.

BELFAST is the capital of Ulster and the name is often used as synonymous with Ulster. This shows BELFAST at a reverse sequence of 81. It begins at 1-Chronicles 4:43 and going upwards ends at 1-Chronicles 5:10. Here are the highlighted names that intersect or lie close to BELFAST. The significance of these names in a Lost Ten Tribes context should be clear to followers of Brit-Am writings.

KING OF JUDAH
THE CHILDREN OF SIMEON
THE REMNANT OF AMALEK
JOSEPH
THE FIRSTBORN OF ISRAEL
THEY WERE WHO EXILED BY
TIGLATHPILESER
IN THE LAND OF GILEAD

These results are quite significant and impressive and cannot be attributed to chance.

Heritage

Yes, laddie, the bones of your ancestors
lie in this land as witness to our faith.
Here came Jeremiah, Baruch, and Zedekiah's daughter,
over the water, to Emerald Isle,
where David's harp changed us all.
We have our music, our legends, our artifacts
—the Nelson stone, the Stone of Scone,
the stone of the Kings at Tara,
symbols of our birthrights going back
to Jacob and his pillow
and even the twelve stones in the Jordan.
The Hebrew in our name
British meaning Covenant Man.
Yes it fell from our forefather's tongues
whatever we've done, wherever we've gone,
Solomon's navy, once anchored at Nuweiba,
went before us.
And is there a one of us that wouldn't fight
to defend the honor of that claim?

Footsteps of the Prodigal

When a fool walks down the road,
he tells everyone he meets that he's a fool.
—Ecclesiastes 10:3

for ye are like unto whited sepulchers,
which indeed appear beautiful outward,
but are within full of dead men's bones..
—Matthew 23:27

Thou hypocrite, first cast the log out of thine own eye; and then thou shalt see more clearly to cast the speck out of thy brother's eye.
—Matthew 7:5

The spirits I summoned up,
I can't now rid myself of.
The Sorcerer's Apprentice,
—J.W. Von Goethe

History, Stephan said, is a nightmare

from which I am trying to awake.

—James Joyce, Ulysses

Humankind cannot bear very much reality.

—T.S. Eliot

Only two things are infinite, the universe and human stupidity, and I'm not sure about the former.

—Albert Einstein

We are inclined to believe those we do not know because they have never deceived us.
The Idler,

—Samuel Johnson

He that is born to be hanged
shall never be drowned.

— Thomas Fuller, Gnomologia

It is better to be Socrates dissatisfied
than a pig satisfied.

—J.S. Mill, The Subjection of Women

Power is the greatest aphrodisiac.

—Henry Kissinger

The Quarry

I went up to a quarry,
abandoned long ago,
way up in the mountains,
where only bedrock rose.
The road was steep and narrow,
on cliffs that took my breath,
and the things I found there
were mostly made of death.

All around the ruins,
boards and nails were strewn;
time was just a foolin'
with remnants of high noon.

I tried to find a reason
for why they all we're gone:
the men who swung their hammers
into evening from the dawn.
But the bones gave no answers,
and the mountains never would
-the silence that surrounds them
cannot be understood.
All around my vision,
dreams are being undreamed;
time has the decision
to make what matters mean.

My name was in that quarry,
abandoned long ago,
forged as it was spoken,
by men who could not know,
how time would tear asunder
the work their lives had been
-although we build then over,
I saw how it must end.

All around our doings,
are visions of high noon,
all on top of ruins,
that time has made too soon.

The One-Armed Man

& Other Obtuse Illustrations

	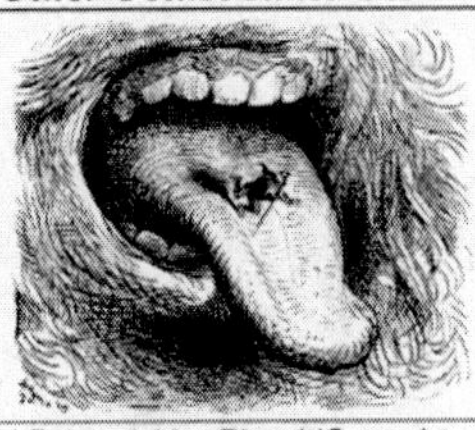	**Poet on last leg** 
Aardvark & Ant	I-Robot, Adrian Ricard (Granny) to Will Smith (Detective Del Spooner):"...***sometimes the stuff that comes out of your mouth..!"***	

Freed Though Amputeed

A poet strains,
transfixing scrawled images
like intense intestines
coil about a gut feeling.
Joy and pain couple beneath his pen,
holding him like a voyeur
when lust steams up a window.
But this cathartic individual
gnaws his wrist in ritual
—letting his grip go,
to be freed though amputeed.

In the gluesome grasp
of an aardvark on an ant,
he is suddenly sucked
from his fatal error:
becoming the stuff that's swallowed
in this age of reader-eat-writer.

In ecology, a niche is a term describing the way of life of a species. Each species is thought to have a separate, unique niche. The ecological niche describes how an organism or population responds to the distribution of resources and competitors (e.g., by growing when resources are abundant, and when predators, parasites and pathogens are scarce) and how it in turn alters those same factors (e.g., limiting access to resources by other organisms, acting as a food source for predators and a consumer of prey). From Wikipedia, Ecological Niche

There is a societal myth that each person fills a particular niche in society and in their workplace. The myth is so prevalent that society produces movies to indoctrinate children into this world view such as ANTS & ANTZ (in ANTZ the human boy child is taught to realize it is He who does not fit in?! And by the way, he needs to worship the Ant Mother??! I think it's Hollywood elite that need to build similar niches!?

Similar Niches

I know too many literal people.
They have certain words they've claimed
to control their conversations.
You can't speak to them
except within their boundaries.
They feel secure in their understanding
of big problems from their own bailiwick.
The world's problems relate fundamentally
to arrangements of atoms
that can be manipulated
by science and sweat in tandem.
But we transform matter into political issues
with paper and passion and nameless factions.
I'm sure big problems could be solved
if there weren't so many corners constructed
in a naturally spherical world.
The trouble is
both space and power
are a function of the number of your allies.
No one can see back to the womb
and few can even imagine tomorrow.
We live in a series of todays,
worrying about close horizons
and regretting yesterdays.
And in our misery
we seek those constructing similar niches.

Retrospective

(Sorry about your divorce, cowgirl)

I remember a day
when my frog died, and my friend lied,
and my dog tried to bite my hindside
Feeling lower than a snail's trail,
and bluer than a thumbnail
under a cross-eyed carpenter's swing,
I ran away from everything.
But I found a new frog 'round the next bend,
and I took that darn dog down to my ex-friend,
who promptly bent over in front of old Rover,
and now is a preacher against the fanged creature.
So I've seen bad times,
turn into better times,
and the face of a former friend,
looks better from the rear end.

Charleton Heston as El Cid

Got an Answer Anyone?

Decades ago, when wars were fought on other continents,
our boys fought *over there*.

We saw gruesome images on the television,

But we had God on our side, and stood strong abroad and at home.

But these days the media is convinced that our faith was empty, our convictions were evil, and we don't have enough homosexuals anywhere.

Now we are so confused; we think Christians are dangerous and Jihad is exaggerated—an artifact of an archaic term not actualized since the walls of Vienna weren't breached, the Hammer pushed them out of France, El Cid vanquished them in Valencia, and the Muslims lost in Granada in 1492, so Al Andalus was lost. General Muhammad said as he sailed away, "Do not cry like a woman for what you could not defend as a man." These days we have Muslims in the Whitehouse, Muslims in the Congress, and the Muslim Brotherhood teaching the military protocols for Muslims under their command. Heck, we financed the Arab Spring. Is there anything more stupid we have left undone? Got an answer anyone? We need a new hero.

Sycamore tree trunk at Ground Zero by St. Paul's church

A hundred year old Sycamore tree was destroyed by a piece of the World Trade Center and protected St. Paul's church.

1 King 16:30-33 "Now Ahab the son of Omri did evil in the sight of the Lord, more than all who were before him.

The region was conquered by the Assyrians in 722 BC, and most of its population was taken into captivity and deported. These were the famous Ten Lost Tribes of Israel.

The Continuing Melodrama of Isaiah 9:10

All our illusions
can be dispelled
in swell foops too close together
to contemplate in a rational manner.
When the north wall collapses
and you can't hide by the south
cause it fell in last week,
it's time to dwell outdoors
where there are no familiar structures
to depend upon.

This is the ruins of Samaria left by Sennacherib.
Looks like Isaiah 9:10 to me.

When rolling with George Bush, every ride begins with a salute. (Photo: Bryan Frazier) President George W. Bush rides 100K Off-road in Palo Verde Canyon 20 with Wounded Warriors

Liberty or Death

I prefer
Patrick Henry's first alternative,
but sometimes
we can only achieve it
by accepting the second.
Nevertheless,
either end is preferable
to appealing the verdict
and repeating the trial.

Been there!
Done That!
Got Some!!

Got Some

I don't depend on walls...
brick and mortar crumble
at predestined moments
regardless of calculated probabilities
of their integrity.
Integrity is something
only rare individuals possess,
while the world
plots their destruction
remorselessly...
The world wants integrity
doled out evenly,
so everyone can say they've got some.

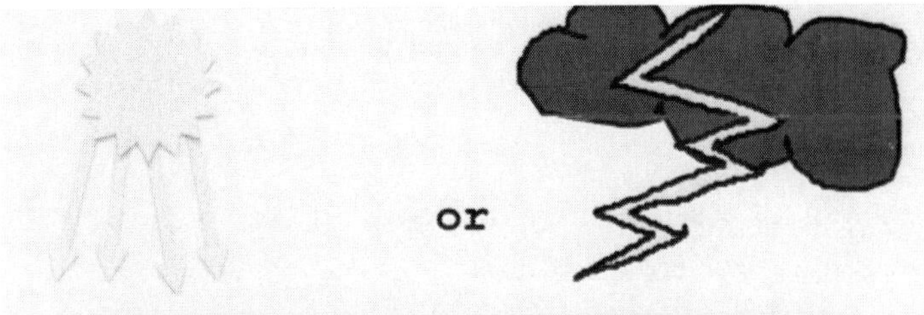

SLOWLY, ALMOST IMPERCEPTIBLY, FEELINGS OF FUTILITY STARTED TO CREEP IN.

Too Many

So many lives
go walking wasted thru dark doors
leading to yesterday's futility,
repeating motions of meaning
before dull eyes disconnected
from the higher heart,
groping for a program protecting them
from the wince
any sunshine or rain can bring.

So many times
go wasted walking to daily depths
of routine at any length.
Buried in automatic response
is a desperate thirst,
unquenched by layers of miscellaneous
meant to still life
with only a gesture of the immortal
in a stream of unconsciousness.

How can the aching heart hurt thru
the dense to and fro
we've built to soothe it?

Your Eyes Yesterday

Once we were younger,
driven by a hunger within
—like the wind is driven by itself.
So far we flung our hearts
to where we've sung our parts of searching songs.
But the lyrics fell faint in the distance
and isms failed to explain our existence
in a void filled only by Maranathas and Mariahs.

Had I known fifteen years ago
that we'd be home for a minute or so,
I'd have rehearsed reunion speeches
and polished the clumsy reaches
that finally held you for a moment.
For in our fleeting meeting
years of seeking meaning
lost their substance,
and I found myself careening
through what was once:
caressing the breast of truth,
clinging to the rest of truth,
we bore the banner of our quest for truth,
but all we really knew was youth....
Yet, I was so surprised
to awaken to your eyes yesterday,
that I fumbled for my emblems anyway.

Generations

Genesis 9:12	And God said, This [is] the token of the covenant which I make between me and you and every living creature that [is] with you, for perpetual generations:
9:13	I do set my [Rain]bow in the cloud, and it shall be for a token of a covenant between me and the earth.
Luk 21:32	Verily I say unto you, This generation shall not pass away till all be fulfilled.
Joel 1:3	Tell ye your children of it, and [let] your children [tell] their children, and their children another generation
Mathew 23:36	Verily I say unto you, All these things shall come upon this generation.
Luk 1:50	And his mercy [is] on them that fear him from generation to generation.
Ecc 1:8	All things [are] full of labour; man cannot utter [it]: the eye is not satisfied with seeing, nor the ear filled with hearing.

Ecc 1:9	The thing that hath been, it [is that] which shall be; and that which is done [is] that which shall be done: and [there is] no new [thing] under the sun.
Ecc 2:14	The wise man's eyes [are] in his head; but the fool walketh in darkness.
Ecc 2:24	[There is] nothing better for a man, [than] that he should eat and drink, and [that] he should make his soul enjoy good in his labour. This also I saw, that it [was] from the hand of God.
Ecc 2:26	For [God] giveth to a man that [is] good in his sight wisdom, and knowledge, and joy: but to the sinner he giveth travail, to gather and to heap up, that he may give to [him that is] good before God. This also [is] vanity and vexation of spirit.
Ecc9:10 [RSV]	Whatever your hand finds to do, do it with your might; for there is no work or thought or knowledge or wisdom in Sheol, to which you are going.

Boy's Dream

She's got eyes from a fable
where the villager's sigh
in ephemeral contentment when she passes by.
She shed tears that enable
even statues to cry,
and sob little stones
that the sunlight won't dry.
Her voice is adorning each early morning
with meadowlark charm,
and her feet on the pathway
will never walk half-way
when her aim is my arm,
and my shoulder would hold her
.....if only I were older.

Mississippi River, below Lock and Dam 14, Iowa Side, Late fall 1981, Bald Eagles are gathering

Talking Old Shoulders

How my shoulders
loved the seat of your pants.
We slowly danced
a thousand miles,
one or two at a time.
Suddenly your anxious feet
wanted their own way.
Now I cannot keep up,
strangely off-balance
behind my beloved burden.

Only Children Know

Children,
fresh from the hand of God,
like the world when he rested,
are laughing mirrors of his love.
Only children,
find beauty in every face they see.
Only children,
know everything was created for their joy.
Looking in the eyes of our children
we touch again our own innocence
—innocence we thought a dim and fading memory,
suddenly remembered with the excitement
only children know.
God it makes a mother glow.

W H E N
H O
Y E S
O
H O W
N
Always

Parent's Dictionary

Why is the question
that must always be asked.

Because is the answer
that will never do.

When is the future
that no one can wait for.

Soon is the time
that isn't soon.

How is the wonder
that fills all the ages.

If is the lever
that moves all our dreams.

No is the tear-maker
that hurts even in
the speaking.

Yes is the gift
to gleeful children's hearts.

Now is the answer
to questions yet unasked.

Always is a word
only children can believe in.

...very dangerous...
you go first.
not too soft a landing spot
PHILOSOPHY
This is humiliating.

Flightpath

My flightpath is too low for birds
—it intersects the earth.

Sore Uncanny Leap

Even Thomas believed his eyes,
why must you intellectualize?
Though you piled it high and deep,
it still don't smell too sweet.
Perhaps I'll keep my feet.
or Soren can I leap?

A Backward Dog

A backward dog is in the dirt.
Beneath the trees, beyond my pleas
for his grin at the door,
he doesn't hurt anymore.
Now he greens the leaves in spring,
and what I know of needs is
everything
grows this way
—from someone in the clay.

Once I looked for God
in a backward dog.
Now I see
in mirrors are the peerers,
and God is not seen, but known.

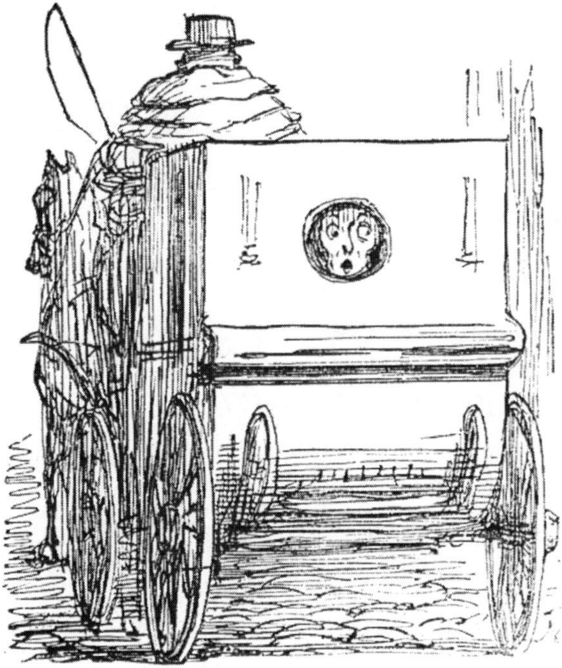

Time Happens

I was nearly clocked last night.
That is, I repeatedly rammed into a wall of time
during an endless iterative evening
—my forehead is flat from interrupted accelerations.

How is it some days are long
and some are short
without the influence of hallucinogens?
There must be some algorithm
to explain such phenomena:
{[(stress)2 + (pressure)3] (log[(metabolic rate)/(endurance)])}
= C1/2
or maybe not.

Time seems to bend when an accident
arrives suddenly from peripheral visions.
Time passes infinitesimally
during a week with the in-laws or a day in court.
Time speeds up incrementally after virginity,
and increases exponentially
with age and under the influence of teenagers.

We all seem to start out
with an equal and infinite supply of time.
Yet time runs out suddenly and prematurely for many
while some seem to have more than they deserve.
Jesus said, "Time and chance happen to all men."
I understand chance in a probabilistic way,
but it takes more than experience
to fathom time *happening* to me.
It wouldn't sell on a Tee Shirt
but it will eventually make me dirt.

Retreat (Moss Haven)

Striving to revise
the big picture
to reflect my own perceptions,
I inadvertently dropped my facade.
Encountering a panorama
my cloistered self couldn't nomenclate,
I retreated via tropism
like moss to the north side.

Exception

If I had time I would bend it around a french curve to suit my inclinations. I'd like to accelerate through days like today, and sashay, dawdle, lolligag, loiter, drowse, drift, and dream through the night. Despite astronomer's theories and the vivid and incredulous imaginings of evolutionists and paleontologists, I am sure that time is an illusion, that space is only a concept, and that the world is not flat, but *maya.* Brigadoon was and will be the only exception.

Dear Prudence

What good are clothes,
as if anyone knows?
All night they're off,
all day they're on,
and out on the lawn
you've got moron.
It seems such a waste,
so utterly chaste,
when you're so cute
in your birthday suit.
But the clothes knows
the closed nose
can only result
if one adult
the cold winds assault.
Over weather together
we might win,
but to endure it alone
is a cold, cold sin.

There are times
when you really feel
a part of everything...

Wedding Song

My heart's clear water, always running,
seeking sister streams by river's reach.
My arms are fair forests, green and growing,
bearing brother beasts in silent speech.
I speak in rain, I shout in wonder.
I see in lightning and in wonder.
I laugh aloud, I cry in vain.
I am the dawn's illumination,
reflected sunlight on the sea.
I am the lifeblood of the ocean,
moved by tides the moon provides,
and it was Almighty God that made the moon,
without which we could not survive,
no matter how we strive. So let's stay alive,
and remember that moonlight
is a mighty fine thing on a wedding night!

Mystery

Things I did when I was a kid
are a part of me that's hid
by history.
Now I walk somewhere amid
tomorrow and today,
with no time for mystery.
Yet I wonder if I'm not still a child,
and I wonder if my father smiled,
and if he came in and gave me a kiss,
when he wondered when I'd wonder this.

Replacements

In a children's crusade
many dragons are slayed,
But replacements are made,
replacements are made.

Phillip Eugene Robinson, started working coal mines in Rolla, Missouri, at the age of 5, WWII Navy Fighter Pilot, RIP, 1995, Union Chapel Cemetery, Iowa

As a Child

I dreamed I spoke with my father last night.
It's been three years now that he's been dead.
We never talked much except last night,
and now I can't remember what he said.
It could be a subconscious whim,
making this memory lost,
cause I never could connect with him
—somehow our wires got crossed.
We just couldn't relate man-to-man,
though occasionally one of us tried.
He found refuge (you get it where you can)
in remembering the child at his side,
and I escaped, as when I was young,
in distance, a sort of resistance,
to emotions you can't overcome.
I'd like to hear what my Dad had to say,
though I expect that deep down I've known,
cause I'm sure we both don't want it this way
between me and children of my own.
I can almost hear him saying...
"Tot, be thankful for what you got",
and I think I need to be praying
to practice what I was taught
as a child.

To Renew

Many songs have been raised
in lament of the days
spent away, spent away
from home.
How they bemoan
learning those lessons
and leaving adolescence,
like all virginity that conceives
is gone, but never leaves,
sad memories.

Grieving over leaving
and yearning for returning
will not forgive the sin
but only hold it in.
Remember...
familiar faces, comfortable chairs,
warm embraces, self-creaking stairs,
are held in the heart the way a mother is,
are days that must depart to be relived.
Home has a door
we exit and enter through,
and something more we are meant to renew.

Zöe (2222)

The moon is never so silver,
so circular,
so central,
so like a prayer in my eyes,
as when it rises in winter...
From the dusk of a gray world
into the intangible,
it marks the imperceptible
passing of short days
in a lengthening life.
The event is so singular,
like the division of pi,
unrepeated in perfection
through all of my memories
or the whole of our geometry,
except I know
the conscience of my race
holds such images
collectively
dearly.
Some moments near you
are like visions
of the infinite.

If I had a Quarter Horse

By: Zoe Anna Robinson

I was sitting by a tree one fine summer day, watching the clouds float away.

All of a sudden my eyes started to get heavy, and I drifted into a dream…

I was walking next to a stream where things aren't what they seem. A quarter horse was suddenly nuzzling my hand, and to my surprise I had an apple in my hand for him. We were instant friends and away I went on his back. As we rode, the seasons changed from summer through fall into winter. In the summer we rode lazily, drifting here and there without a care. We splashed like fish through the stream, and we flew like birds through the meadows. Then the mornings grew crisp and the leaves turned crimson. We cantered over the hills, while nature painted them before our eyes, and the echo of my great horse's hoof beats sounded like thunder, saying, "So long," to summer. Snowflakes began to fall and we found ourselves in the deep pine woods. The sun went down and the moon came up. A lonely owl flew across the sky. I would have wondered why, or even started to cry, if not for my great brave friend beneath me. But instead, I thought riding together, or being so much alone, was the best part of my dream. But as soon as I realized it was a dream, I woke up. There were the clouds above I was watching when I fell asleep. But they weren't as lovely now that my great friend was gone. I wrote this poem to try and bring my great friend back a little of his memory…

> If I had a Quarter Horse
> I would surely ride the course
> Toward winter…
> In the woods, in the dusk, in the snow,
> Where only dreamers dare to go.
> My pony's breath would fog the air,
> And I would feel so unaware
> of anything but tracks in the snow,
> and all the places yet to go!

Non Illegitimi
Carborundum
Don't let the Bastards get you down

Non Illegitimi Carborundum

When a fool walks down the road,
he tells everyone he meets that he's a fool.
—Ecclesiastes 10:3

...for ye are like unto whited sepulchres,
which indeed appear beautiful outward,
but are within full of dead men's bones..
—Matthew 23:27

Thou hypocrite, first cast the log out of thine own
eye; and then thou shalt see more clearly
to cast the speck out of thy brother's eye.
—Matthew 7:5

The spirits I summoned up,
I can't now rid myself of.
—J.W. Von Goethe, The Sorcerer's Apprentice

History, Stephan said, is a nightmare
from which I am trying to awake.
— James Joyce, Ulysses

Humankind cannot bear very much reality.
—T.S. Eliot

Only two things are infinite, the universe and human
stupidity, and I'm not sure about the former.
—Albert Einstein

> We are inclined to believe those we do not know because they have never deceived us.
>
> —Samuel Johnson, The Idler

> He that is born to be hanged
> shall never be drowned.
>
> —Thomas Fuller, Gnomologia

> It is better to be Socrates dissatisfied
> than a pig satisfied.
>
> —J.S. Mill

> The Subjection of Women
> Power is the greatest aphrodisiac.
>
> —Henry Kissinger

> And this [is] the writing that was written, MENE, MENE, TEKEL, UPHARSIN
>
> —Daniel 5:25

> This [is] the interpretation of the thing: Mene, Mene, Tekel Upharsin;Thous art weighed in the balances and art found wanting.
>
> —Daniel 5:26

And you know, it could happen again, here and now.

There is another Babylon predicted and we are about as deserving as could be expected before fire and brimstone fall.

With Some Regret

The handwriting's on the wall,
and the clock just stopped.
How unsettling the step
from time to eternity,
the realization...
all you thought was solid is without substance,
and what has substance is so unexpected,
but so instantly understood

as the manifestation
of childhood memories,
dismissed long ago
by ego seeking emancipation
as mere superstition.
Existential meaning is now anachronistic,
and the truth you always said you sought
stands where you want it not,
and there is no epithet
you can utter
that won't be hollow,
and there is no saliva
that you can swallow,
and there is no place
you can hide your face,
and there is no end to this embrace,
cause there is no time or even space
that you could measure at your leisure,
cause there is no leisure,
no more pleasure, no more treasure.
What you see is what you get,
with some regret.

Politics at the Speed of Light

I've seen a hundred people
say ten thousand words
and be five hundred nerds
as they milled about
one man without doubt.
Showing all their faces
and saying greater graces
as they sought
concurring groups
not outnumbered by opposing troops.
They eventually united to crush him:
the evolution of slogans and euphemisms
was impressive,
but they never would have won
without a character assassin to lead them.

Herding Porcupine in Jemez

Truth Through Adjectives

I hear that Madison Avenue
has hired Tom Sawyer's spurious offspring
to trick the multitudes
into white-washing their sepulchres
with every article of toilette
on sale.
However,
I have no consuming urges
while driving rush hour freeways:
It's like having to herd
a sedentary porcupine
through a maze with your nose
while being followed by
a perverted unicorn,
so I damn sure turn my radio off.

..then bring me my favorite camel hair gloves, and be quick about it!
?? I don't..

He Gloves Me Not

In this town are many men
that my boss stands above.
He reaches down and touches them
with his tender glove.

The glove is his compassion
for his fellow man.
He loves it for the fashion
that it fits upon his hand.

A part of his heart
is the stuff of that glove,
the very hair of his head is the thread.
But the eye of the needle
was made too small,
so the glove that he loves
isn't there at all.

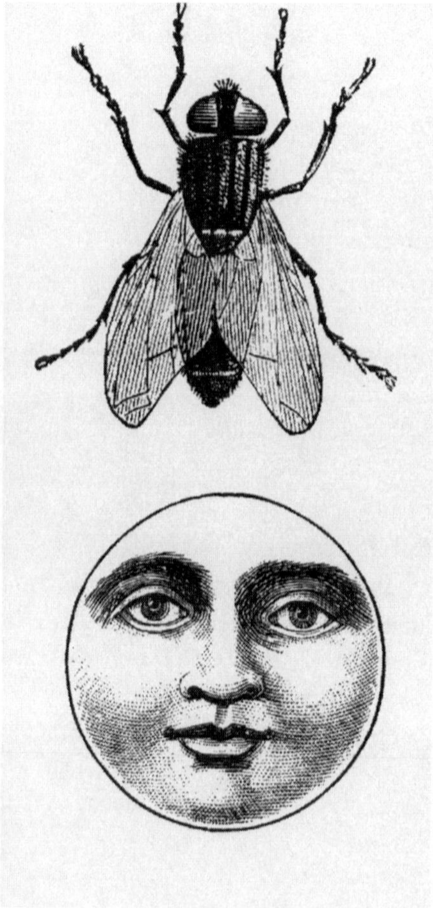

The Big Black Bug has Black Blood.
Say it ten times fast, I dare you!
Then swat him!!

Flies Love Garbage

Flies love garbage.
You've seen them
—rubbing their hands together
in anticipation.
Unfortunately,
the world is cursed
with nearsighted flies
who give everyone
personal scrutiny.
Lucky for flies
—it all looks the same
to them.

UPI:	dingos attacking townspeople in western Australia....
The Monitor:	Third coyote attack this year is troubling....
AP:	Rabies epidemic is eastern U.S....
UPI:	Wild dog packs in Dakotas threaten livestock...
Albuquerque Journal:	Ranchers protest reintroduction of wolves...
Miami Herald:	Alligator attacks on the increase as species proliferates.....
TLC:	5 Fatal Mountain Lion attacks in 3 years; multiple near misses; Bear attacks rise

I would pay attention to the beasts of the earth (see Rev 6:8.)
The earth doesn't really need saving, but you may....

Backyard Dogs

(Revelation 6:8)

If they're good dogs
(And everyone thinks his is),
one dog is a delight,
two dogs can be a little trying,
three dogs are a nuisance,
four get out of hand and eat the neighbor's chickens,
five are a pack with a pack mentality,
and six are uncontrollable without a gun.

Convicts say missing five meals
will make any man a thief.
Dogs that miss five meals bare their teeth
at the hand that feed them,
and what follows is the bedlam
Sci-Fi writers make unreal.

How many dogs are there in your neighborhood,
and how do you suppose they'll act
if the dog food ever runs out?
Are there enough Vietnamese in your town to eat them,
or are there enough dogs to eat the Vietnamese?

And who do you suppose they'll eat next?
Evolutionists would assert,
only those least important to the survival of the species...
they should stay alert,
cause dogs often find prey by scenting their feces.

Factories

Factories were made for the Japanese:
they seem to love to drone together,
singing company songs,
and making robots for their ancestors.
I delivered a vacuum breaker intended
for downcombers in the reactor building
of a Japanese power plant.
They gave me Mitsubishi cufflinks
on December 7 for my honorable service.

Factories kill Americans:
smother them with sameness,
make them small and malleable,
even easily fashioned.
Factory owners love them though,
admiring the Japanese.

There
must be more
money, there
must be more
money
(sigh)
Radix
Mallorum
Est
Cupiditas
NEON

The Neon of Desire

A subtle creature
is difficult to transfix:
a rocking horse preacher
must intermix
riding and pleasure,
and who can measure
the parts of the sum?
They are indistinct
except in the neon of desire
that shines from within
kindred spirits.

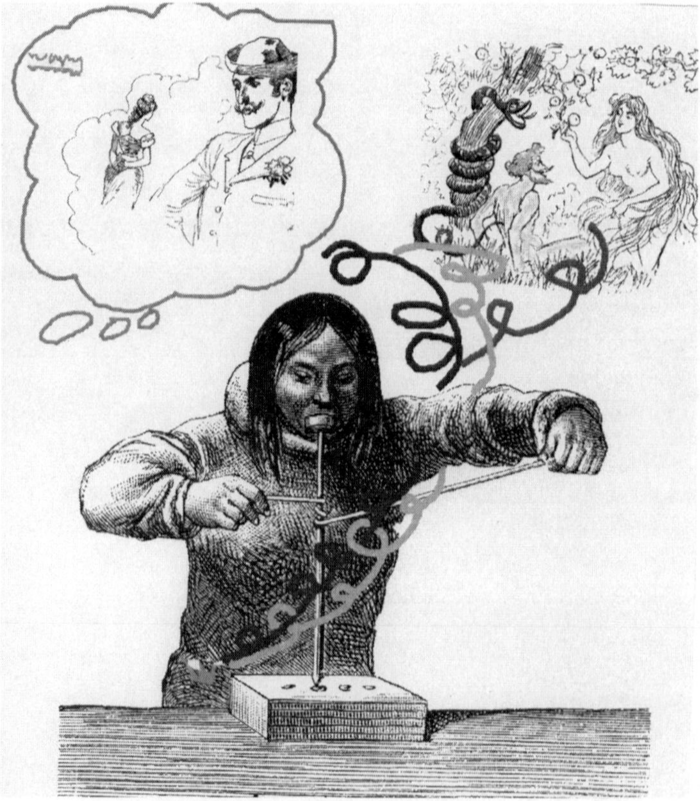

Fire in Our Fantasy*

Sunlight bursts transluced her skirts,
surrounding her silhouette at first
with a shimmer, and then with his imagination.
Favorite flirts that they rehearsed
became daydreams, nestled and nursed,
from a glimmer, to an inner conflagration.

Her chameleon eyes completed her clothes,
and beneath them, Heaven knows,
he longed to explore.
But there our whys and all our woes
are undressed like Eve's old beaux,
who failed to adore
innocence over desire,
who founded the fire
that burns in our fantasies.

It's so sad when two fall into flame,
and many you knew will do the same.
In these last and driest days,
desire sets the world ablaze.

Via Lares Deo

As I walked out on the streets of Laredo,
it was best to wear a vest and stay way low.
Everywhere and every day,
there's a gun going off

in a virtual ritual way:
pulling a trigger,
like squeezing a remote,
brings an instant end
to adversary and friend,
and changes situations like surfing channels.

Children shoot children,
in and out of school,
and pronounce it cool;
gangs shoot other gangs
and anyone intersecting their aim,
while speaking secret slangs,
and questing for a name;
siblings shoot parents,
mothers shoot fathers,
fathers shoot families;
Muslims shoot anyone
who doesn't bow to Allah,
and many who do,
but mind another mullah;
the Croats, Belfastians, and Hutu,
xenocrats, neonazians, and Jews too,
learn to slay like their parents slew...
and of course,
agents shoot the elusive
—especially the reclusive;

And isn't it a shame
the world's gone so insane
in every direction
from the streets of Laredo,
where the showdown
between good and evil
is now a meaningless anachronism,
conceptually foreign
to the majority of shooters
that have no connection to history
beyond their unconscious embodiment of visions from another age.

Yes, the boyz in the 'hood
and the girls that would
and mothers that must
and fathers that lust
and those that just
hate others born of dust,
all kill for convenience
in an age of expedience.

And of course,
where there's a warrant,
resistance is abhorrent,
and deadly force is legal
—sometimes even regal.

Tomb of Kar

Dr. Zahi Hawass, who is the head of The Supreme Council of Antiquities in Egypt and known throughout the world of Archeology, remarked that the image of Kar looked just like President Obama on a trip to the pyramids at Giza in May of 2009.

Kar lived about 4,600 years ago in ancient Egypt. Kar was a great and devoted Black–African (Khemitian) who had served one of the Pharaohs of Egypt. Just like President Barack, Father Kar was a priest, a scholar, a lawyer, and an advocate. He later became a great judge.

Barack Obama: "That looks like me, look at those ears."

Truth without Threat

and Threats without Truth

Decades ago
when wars were fought on other continents,
our boys fought over there.
We saw the gruesome images on the television,
but we had God on our side
and stood strong abroad and at home.
But these days the media is convinced
that our faith was empty, and our convictions were evil.
Now were so confused we think Christians are dangerous
and Jihad is exaggerated, so we send our hard-earned dollars
to support the Arab Spring
like we somehow owe a Jizya tax.
And we pretend Muslims are right about everything.
Already deluded financial institutions are setting aside the
Zakat and pretending it is a good thing.
Not to me, maybe to you. Boohoo.

Melodrama

All our illusions can be dispelled
in *swell foops* too close together
to contemplate in a rational manner.
When the north wall collapses
and you can't hide by the south
cause it fell in last week,
it's time to dwell outdoors
where there are no familiar structures
to depend upon.

The New Censors

How we preach against intolerance.
We must have diversity of thought, speech,
sexual persuasion, ethnicity, and spirituality.
But don't smoke in our presence
or dare to disable an airbag.
And you must understand...
conservative ideas are intolerant,
and Western civilization is decadent,

and pro-life groups shouldn't be allowed to protest,
and what we define as hate speech is a crime.
Monogamy is an outdated concept and
preaching abstinence infringes on individual rights.
And some groups are more equal than others
and deserve special rights.
We need to bring down dominating white males
and marginalize Hasidic Jews
and make everyone understand
that fundamentalist Christianity,
with its premise of a jealous creator God,
is so anthropomorphic.

We shall overcome
by driving conservatives from the faculty,
by proselytizing the children
with condoms and cartoons,
by failing creationists in every classroom,
by litigating every Christian
and/or heterosexual organization,
by rewriting history,
by legislating diversity.

Just try to get a job Mr. Jones.
We dare you to pray in a public place.
Just argue with evolution and see what grade it gets you.
We've got your number.

...down the hall, just past the Fetal Inconvenience Clinic, across from Euthanasia Care. If you come to the Neonate Elimination Center or the Cloning 2nd-Class Citizens Laboratory,

Follow the Morning Star

The Germany of the 30s
started with euthanasia for the old and infirm
—one doctor recorded the demise of his wife,
poor suffering multiple sclerosis victim;
he tenderly gave her an overdose
to the strains of Wagner.
Those doctors built the first gas chambers
(and ovens to destroy the *defective* bodies)
to dispatch with mercy and German efficiency.
Later the Nazis turned those tools of mercy
to the elimination of the inferior
—the retarded, the Gypsies, and the Jews.

We started with the unwanted.
We proceeded to the less desirable.
Our eugenics were more sophisticated,
quieter, politically sensitive, but no less ruthless.
We abort because children are the wrong sex,
the smaller of twins have Down's Syndrome,
some suspicious gene, or were just too inconvenient.
Who do we abort next, and
why not terminate neonates?

We stick our eggs in surrogates.
We freeze them for later use,

and the option of freezing the fertilized
is the stuff of medical journals.
Lesbians demand the sperm of their choice
to make little likenesses.
Hollywood has already predicated Schwartzenegger
as a surrogate mother, pandering to similar gay aspirations.

Everyone knows gene modification and cloning are coming.
We are so enlightened, so smug,
so capable of anything we imagine.
But we are not God, or even little gods
—we are just tinkering with designs we don't understand.
On the edge of a precipice, we refuse to recognize,
though many have strode here confidently in days gone by.
Does anyone remember Molech?
Does anyone know who his father was?

OK

Diversity
Sexual Persuasion
Eastern Religions
Sharia
Gaia
Wiccans
Wolves

Not Ok

Evangelical Christians
Orthodox Jews & Israelis
Conservatives
Single White Males
Ranchers
Purveyors of Raw Milk
Real Scientists
Nationalism

Have you seen this man? His Name is Nimrod and he's looking to enjoy running the world after being unavailable for 6000 years.

E Pluribus Unum et Vinculum Vobiscum

Sages of the ages
longed to see our day,
and did not.
These days sages
make minimum wages
cause they're not sought
to right our way.
In our race toward oneness and *talkspeak,*
leaders rise like curds
and spread soft words
like manure
on ignorant lands in the midst of intellectual blight
—Oh fer sure,
no one understands,
but we all have something to hear and recite.
Law proliferates
and seems like it liberates
as it forbids what discriminates
against litigated license
(and some symbolic bisons),
and we, borne on a flood of arrogance and avarice,
rush to rule the world (or what has become of us)
to benefit as many as can be,
on board with a Nimrod wannabee,
without rudder or compass, but at the helm,

preparing to number us, and overwhelm
any remaining reactionary,
as we draw seemingly unitary
like a barrel to the falls
—enamored of the calls
for something more extreme,
believing what evolves
a self-fulfilling dream.

Mirages

I've got fleas on my knees
from crawling toward mirages,
and frostbite on fingers
from meeting Mr. Rogers
—still it's nice to be my old neighborhood
after dreaming for years how it would.

DM
DREAMS

Departure

While I was dreaming of woodstoves and windmills
I became a scientist, and imagine this...
while I yearned for green hills through windowsills,
I rose from obscurity in matters of national security.
Yes, I worked hard, as it was my ethic,
but clearly, God used me in spite of the epic,
I still hold as my glory...
You see, in me lies a story of a reclusive renaissance man,
a learner, a discerner, and a writer because I can
—no, because I must—
live simple and self-sufficient by the work of my hand,
and leave a legacy of literature I hope you understand,
when I return to the dust.
But it seems my daydreams have not come to pass,
and yet the path gleams before me whenever I ask
for guidance in accomplishing the next incremental task.
Who knows what I'll do tomorrow without really trying—
only the Almighty, there is no denying.
It is said that God gives us the desires of our hearts.
in my case God saves me from the vanity a dream life imparts,
and manages to use me when such illusion departs.

Vox Humana

...ye shall be as gods, knowing good and evil..	Genesis 3:5
Who told thee that thou wast naked?	Genesis 3:11
In the sweat of thy face shalt thou eat bread, *till thou return unto the ground, for out of it* *wast thou taken, and unto dust shalt thou return.*	Genesis 7:19
For we know that the whole creation groaneth and travaileth in pain together until now.	Romans 8:22
If I had a dollar for everything I've done, *there'd be a mountain of money...*	*Missionary Man,* Annie Lennox

My karma ran over your dogma.	Unknown
I am a parcel of vain strivings tied by a chance bond together.	*Sic Victa*, H.D. Thoreau
The eye is not satisfied with seeing,, nor the ear full of hearing.	Ecclesiates 1:8
There is no cure for birth or death except to try to enjoy the interval.	George Santayana
Strife is better than loneliness.	Irish Proverb
In a fight between you and the world, bet on the world.	Franz Kafka
In this world ye shall have tribulation. Fear not, I have overcome the world.	John 16:33
Sometimes a cigar is just a cigar.	Sigmund Freud
Nothing gold can stay	Robert Frost
Death sneaks up on you like a wind-shield on a bug.	Unknown

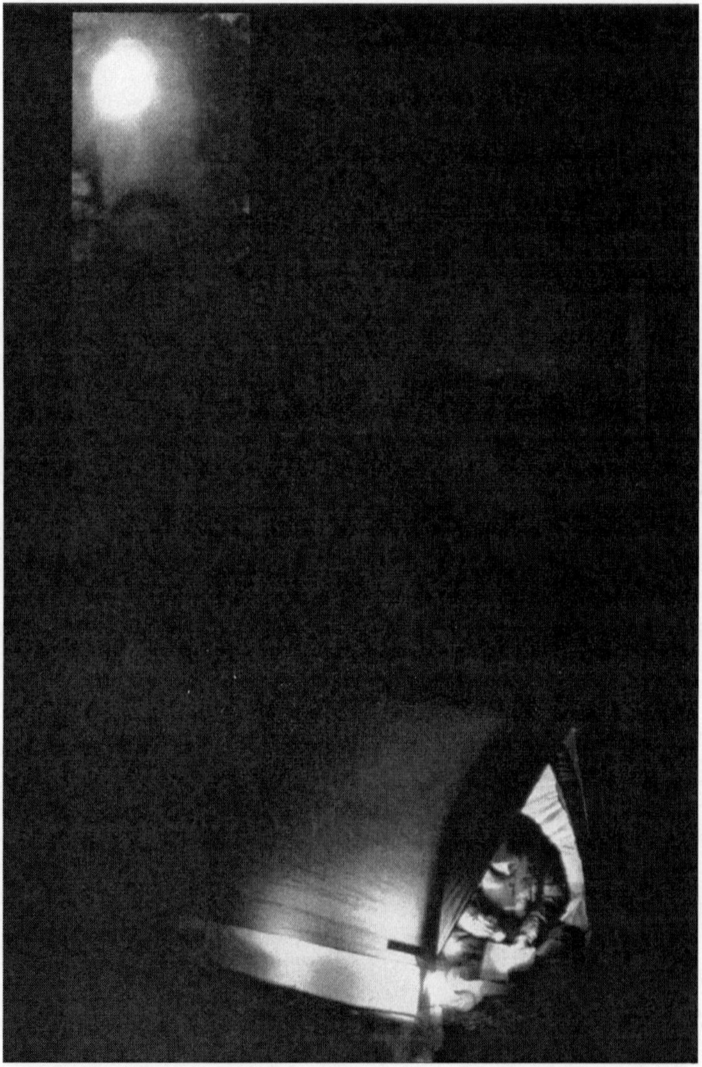

I Think I Can Still Rite

Name me something frivolous
and I'll see it solemn.
Show me someone posthumous
and I'll try to call them back.....
Though they're invisible now,
I'm pretty certain they're black
in the casket under the make-up.
And I'm confident it's black
underground where we put them to rest.
Black is all colors combined and all history underlined and
all mystery of the mind
that remains entwined with rites of mankind
...let's not go there.

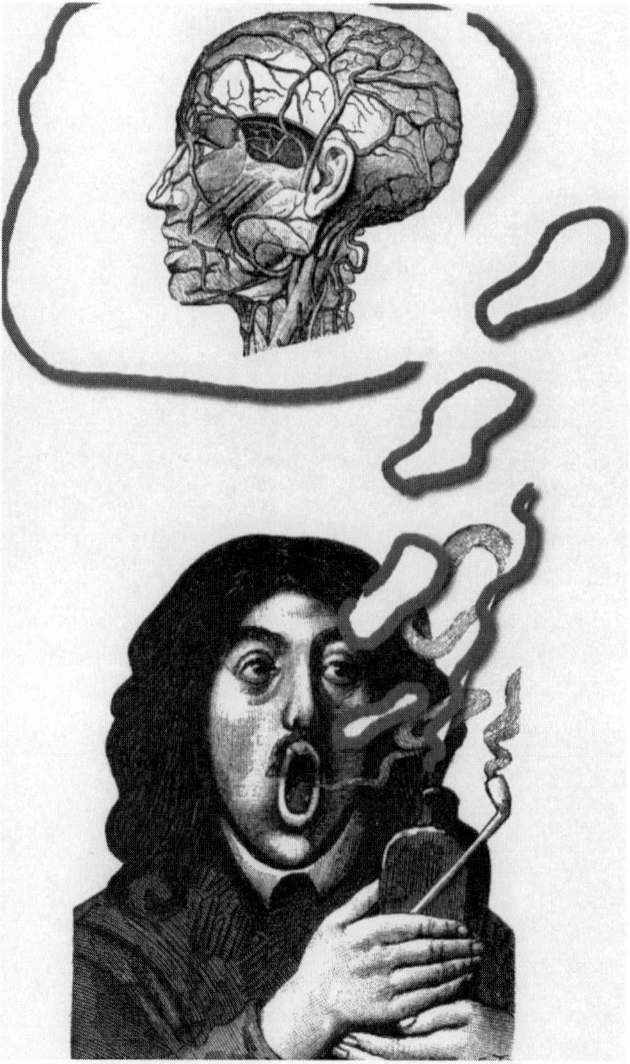

My Own Volition

Buried deep within my chromosomes
are all truths and falsehoods
germane to my race,
and the expressions of my face
are their cumulative response
to today's exigencies.
Psychologists dissect them
by cause and effect,
and philosophers expect them
to logically connect
by means of intellect,
but I just accept them...
How could I reject them?
By my own volition?
Not me, Zarathustra!

"Was ist los?
Ach, ya...
die *itch*
macht nichts,
ist
pedal-philia,
oder *pedal*
envy, oder
pedal
. . .
Why don't
you shut off
the TV...?

Far, Far, Freud

When my wife has an itch, she scratches my foot.
Freud would say such a switch had a sexual root.
He'd be far from right and far from wrong.

The Partial Awakening

Cowboy Pete felt complete
with just a few possessions,
though to some they had become
mere western obsessions.

There was his clock radio
that got him up for the rodeo
after howling *yodeladio* most the night,
and his kitchen chair
where he sat in despair,
straining to sip some coffee with all his might.
But it took all his strength to open his eyes,
and to his surprise,
he couldn't lift the length of his cup to his face,
though he tried in disgrace.

Sitting there in his long underwear
that he'd worn since he was just strange,
he realized his lifestyle
that he'd had for a long while,
possibly needed some change.
He dropped his chin in his cup
and sucked caffeine up
till he was filled with resolve:
I'll put boxer shorts on
and sleep out on the lawn,
though that'll likely involve freezing my fleas
and I'll probably evolve some hair on my knees.
But I'll be out like a light and sleep through the night
till I can stand like a man and drink coffee from a can!
Might even hitch up my saddle and try lookin' lonesome,
or pick up a paddle and canoe till I'm loathsome.
In time that's what most cowboys do,
if the shows on the TV are true.

What is really unclear is where Pete got the beer
to lubricate his plans and add to the cans
in his collection outside the back door,
when his favorite refrigerator was empty of motivator,
and to go to the store was an incredible chore,
but when he makes his mind up,
who knows where he'll wind up.

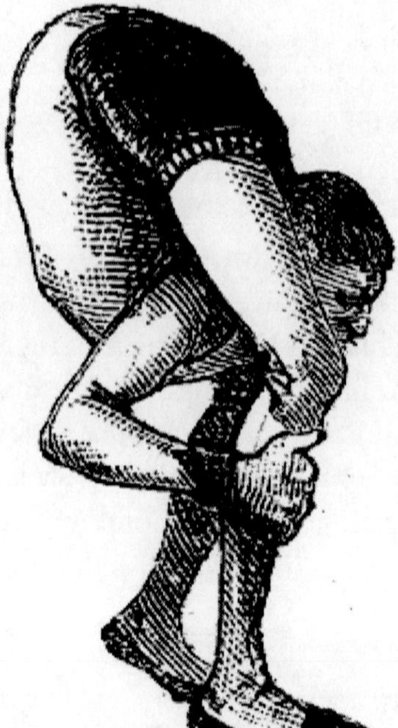

My Goose is Cooked

(by a lesser descendant of Johann Hus)

Sometimes verse is just too terse,
unnatural as buffalo in reverse.
as forced as borscht
down all but Tartars,
as forced as gorgon
on phobic barbers,
as forced as food
toward a snake,
as forced as mood
in a fake,
as forced as excrement
that runs uphill,
as forced as a horse
that swallowed a pill,*
as forced as forth
after poor Sally,
as forced as Valderi
upon a valley,
as forced as sheep
upon a shepherd,
and from 'Bo Peep
(or so I've heard),
as forced as harems
on a eunuch,

as forced as cave canem's
for the dogmatic,
as forced as tits
through a wringer,
as forced as Sphinx
to malinger,
as forced as spoils
from the vanquished
as forced as smiles
from the anguished,
as forced as golfballs
on a course,
as forced as U-Hauls
in divorce,
as forced as apology
without remorse,
as forced as North
upon the Norse,
as forced as payments
from the fined,
as forced as visions
in the blind,
as forced as an opera
for the auditory.
Yes, as forced as a suppository
and similarly anticipatory:
as tired as boats against the tide,
as trite as poems from the fried.

Interpretive Eyes

To all the victims of my imagination,
forced to play roles
assigned by my whim,
born of imperfect perception:
consider yourselves emancipated;
you have until an eventual
mood swing
to assert yourselves,
fashion new identities,
before my interpretive eyes.

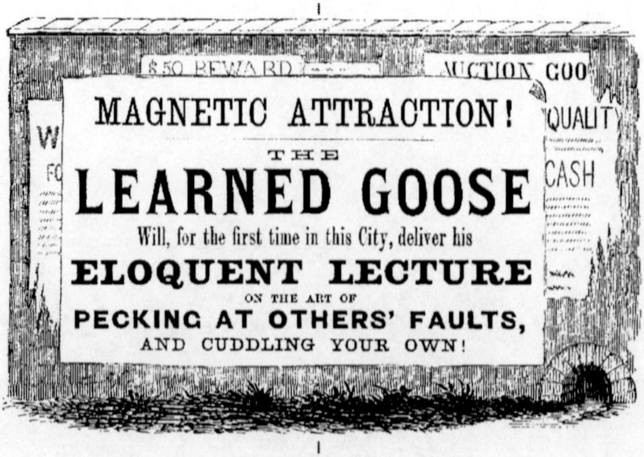

Rays of Light

I looked in someone's eyes tonight
—now I'll never write of sight
in ways that poets do.
Words can repress truth
but words can't express truth
like light rays twixt two.

Inside Strangers

Progressive amnesia is a strange disease:
I remember and forget
the person I used to be,
and rejoice and regret
over old memories.
The symptoms are salty
like tears on my lips,
but the seizures grow seldom
as time takes his tips
—gratuities wasted on declining service.
But the enlightened and lightless,
the frightened and sightless,
are universally retired
by the roles that face them each morning
like weary plastic surgeons.
Suddenly, I don't know the man in the mirror
—and I was always uncomfortable
inside strangers.

Tears of Time

I am filled with the ashes
of a lifetime I burned
with the careless fire
of freedom to waste.

I am poured out
like the wine glass
of yesterday's illusions,
empty of tomorrows.

I am covered with the rags
of daily deceit
I wore a plumes
from Neverland victories.

I am naked
before my leaking eyes
the tears of time
have clouded and cleared.

D
D is anthropomorphic

Distance

Oh Lord,
in this moment
let us realize
how lonely we are for you.
We have hidden our hearts
beneath daily routines
and cares we magnify
to fill our horizons.
For just this moment
let us perceive the ache we make
by filling our thoughts
with dreams of wealth
and nightmares of poverty
and waves of worry beating upon shores
of self-induced euphoria
where we seek to escape.
Only our imagination
gives substance to these vanities,
and in one way:
in the distance we create
between ourselves and you.

The Megacosm

It's not hard to grasp
the megacosm
    from here in
New Mexico:
    there's a handle
on every side.
    The hazard lies in
the slip of the grip:
    Though were both
four-dimensional,
    Stasis is unknown
to it's fluid veneer.

El Megacosm

No es dificil para asimientar
el megacosm de aquí
en Nuevo Mexico:
    Hay mango en cada lado.
    El peligro esta en
el resbalo del aggaro
    Tanto la tierra como la
vida estan encantadas de ese
modo son cuatridimensional,
    Asi es que siempre se
desconooce la chapa de
los movimientos fluidos

    Ay,Yi,Yi, Yi!
(Ay Dios Mio!)

Defector: Suitcase Weapons deployed in U.S... Lebed says 100 Nuclear Mines are missing; Palestinians have 3...Iran has nuclear weapons from Khazakstan...Muslim Bomb is here, Pakistan detonates nuclear weapon...Ezekial 38/39 describes a future intercontinental nuclear exchange...
As the great philosopher,
Sting, said and sang,
"They build machines
that they can't control,
and bury the waste
in a great big hole..."
We must Green the planet!
No Nukes!
We can get everything we
need from the sun!

OOSOOM

I'm tied up in knots over plutonium.
So are most educated men.
Once we thought it was a kind of *utopium*,
but education was different then...

While we were preoccupied,
a generation of Liberal Arts majors
took command of the classrooms.
Serene in their confidence in untenable dogma,
and hysterical in their hatred of
anthropomorphic technologies,
they compassed curricula to pass proselytes.
How completely they graduated the media:
song and dance and twenty-four-hour news
all espouse the same agendas
through single syllables and shocking scenes,
meant to titillate and terrify and tribalize
those with short attention spans.

I usually have no time or strength
to consider this evolution.
I am given to the doing side,
maintaining the deterrent for which
the thoughtful are thankful,
making power that is clean, criticized,
and taken for granted,

insuring the actinide is repurified and reused
—not thrown away where it wouldn't decay
for longer than we have perspective to imagine.
Yes, resisting the headlong rush to the hole
where many would cast all things nuclear,
demanding to be again oblivious
of the little they know of atoms.
Small comfort,
there is always another demon,
and technology is not a box you can weld from the inside,
and the man with destiny's trigger finger
waits in the wings,
drooling.

Our Best Destiny

The approaching heart
accelerates
beyond our reaction time;
without warning the distant
is omnipresent,
overwhelming our perceptions
like continuous epiphany.

Few there are
that meet a heart
with the reciprocate onrush
essential to friction welding
—yes, even fusion,
though fusion is known
as our best destiny.
Be wary of the window…
suddenly light rays of love
can penetrate;
you've only an instant
to absorb or reflect,
as you choose.

Foregone Conclusions

I have always assumed that the last word begins with a *Z*, and that the last winter is either over or will be identified by the absence of spring, and that the last liar will be met with ambivalence, and that it won't matter if the last invitation is RSVP, and I am sure the last sentence will be succinct. I suppose the last whiner will be lost in a crowd. I expect the last moment to be momentous. I suspect the last sex will not be eternally fulfilling. I know the last breath will be out.

Fiat Lux ...Fiat Voluntas Tua

The End

Of making many books there is no end,
and much study is a weariness of the flesh,
but I must confess I want my end
to evoke something stirring, something fresh,
in your consciousness.
It was a carpenter who said, "All is vanity,"
and I've recognized this truth as I sat enthroned near one,
and also when I've seen error in the ways of a dear one
(which is easy when you're near one).
So why is the human condition so sublime and ridiculous?
We sample soufflé and emit the gastriculous,
we make "combustibles" of the deciduous,
we stand in chaos to be meticulous,
and we think "the best" is just a few of us,
yet we care to claim that we're all Spartacus.
We run so hard between Points A and B,
dreaming all the while of what's beyond C,
and rarely remembering who we used to be.
God put the world in our hearts
and set forth the law of entropy,
so though we strive from many starts,
we decline over time like a battery.
Laws of the universe compass our dreams,
but we can see God when he intervenes.
That's why this book is full of such scenes.

Outside the Treasury at the Nabatean Kingdom
of Petra in Jordan, Winter 1994

The Darkness is Deepest in the Light of Day

I tipped a glass with a Saudi billionaire in a bar in Amman.
He flew into Jordan for a drink and foreign conversations,
Hoping to revisit youth and its pursuits in the new Babylon
when he was oblivious and far, far west of Wahabi expectations.
And I smoked a *hookah* with a dozen Hasmonean officials,

on a beach in Aqaba, after midnight, in the moonlight,
where no one would espy their teenage coevals,
manifesting again through synapses awash in nicotine,
and prancing, for the moment, like the Pan's truant shadow.
But in the light of day, under the scrutiny of their peers
and servants of Wazirs, none of them knew me,
except as an infidel…they were nervous in my presence,
counting worry beads as if there were not enough prayers
to cover our association and their intemperance.
Sometimes in the evening we spoke of *Isa* (Jesus),
whom they admire, but don't understand (as in any other land).
But the conversations always seemed to grieve us,
ending on the edge of rage that rises when certainties waver,
when epiphany is juxtaposed with inalienable belief.
Today, rage has grown as thick as smoke in their coffee houses; it's there when good Muslims recite the Hadith,
and explodes like fricatives through clenched teeth.
Yes, rage is the essence of all Dar al Salaam espouses
at the end of the age, and in the light of day.

Originally published in The Best Poems and Poets of 2005, International Library of Poetry

Post Script: Understanding the Role of the fishers, Pagan Beliefs and Practices, and their Influence on Christianity

In consideration of what was planned long ago, and what the future may hold, as discussed in Jeremiah 16:16:

> Behold, I will send for many fishers, saith the LORD, and they shall fish them; and I will send for many hunters, and they shall hunt them from every mountain and from every hill and out of the holes of the rocks.

The *fishers* were sent a couple thousand years ago. That is to say, our Lord, Ya'Shua Ha' Meshiach (Hebrew for Messiah), knew exactly what he was saying in Matthew 4:19, "Follow me and I will make you Fisher's of men." And he certainly chose actual Fishermen as Disciples to reinforce the scriptural connection. The reader will see that the disciples made it their business to locate the Descendants of the Lost Tribes (see next page) and Paul (Rabbi Shaul) made it his mission to find and convert members of the Ten Tribes.

Jeremiah 16 is a record of the Lord banishing the Ten Northern Tribes of Israel from the promised land because they turned to pagan gods and committed terrible sins such as sacrificing their children to Semiramis, wife of Nimrod, called the Queen of heaven (and later called *Ishtar*). Allow me to clarify the significance here by interpretation of ancient pagan history. After Nimrod was killed by Esau, Semiramis claimed Nimrod was raised up to heaven as the sun god and that she was impregnated by the rays of the sun. The supposed son of Semiramus and Nimrod, Tammuz, was an avid boar hunter, and was killed by a wild boar—hence the ancient practice of eating ham on Ishtar (Easter) *Sun*day. She also claimed that she fell from Heaven landing in the Euphrates River and turned a bird into an egg-laying rabbit. This miracle occurred at the vernal equinox, which was then called Ishtar (later known as Easter). This was the origin of a sex cult that performed copulation ceremonies in the fall so that babies could be sacrificed on Easter. Originally Easter eggs were dipped in the blood of these babies…this is the origin of the modern practice of coloring Easter eggs. Some Russian and Greek Orthodox churches still exclusively use blood red dye for Easter eggs today, without knowledge of the origin of this Pagan practice. It was Constantine, a follower of the pagan god, *Sol Invictus Mithra* (a sun god like Nimrod), who brought Easter into Christianity.

The disciples of Jesus clearly kept the Passover as mandated in the twenty-first chapter of Leviticus, as Israel had since the Lord God commanded it as a statute forever in the Exodus. The bottom line here is that the Ten northern tribes were carried away by the Assyrians. Clearly the Apostle Peter and John knew where some of the descendant of the Northern Tribes were, as they mentioned them in letters (epistles):

1st Peter 1:1 and 1:2,

> Peter, an apostle of Jesus Christ, to the strangers scattered throughout Pontus, Galatia, Cappadocia, Asia, and Bithynia, Elect according to the foreknowledge of God the Father, through sanctification of the Spirit, unto obedience and sprinkling of the blood of Jesus Christ: Grace unto you, and peace, be multiplied.

Also, most don't recognize that Caiaphas, the high priest of Israel at the time of Ya'Shua's crucifixion also knew that the Ten Northern tribes still remained viable although dispersed to different areas. His understanding is clear in John 11:49–11:52.

John 11:49

> And one of them, [named] Caiaphas, being the high priest that same year, said unto them, "Ye know nothing at all."

John 11:50

> Nor consider that it is expedient for us, that one man should die for the people, and that the whole nation perish not.

John 11:51

> And this spake he not of himself: but being high priest that year, he prophesied that Jesus should die for that nation;

John 11:52

> And not for that nation only, but that also HE (Ya'Shua) should gather together in one the children of God that were scattered abroad.

Note that according to Hebrew history and anthropology scholar Yair Davidy, the Ten tribes became the ruling clans of the Parthians who held off the mighty Roman armies for centuries, but dispersed into smaller groups when the Parthians were finally defeated by Rome.

Recognizing the Messiah in the Tanakh (Old Testament)

I have spent considerable time thinking about how to respond to questions from Israeli Jews regarding where one finds evidence that the Ya'Shua of the New Testament is the Messiah of the Old Testament. I have decided (1) that I would like to respond in a very straight-forward simple way, and (2) I am compelled to respond in depth regarding God's purposes from the end to the beginning. By the way, I hope I don't offend you in using this term for the Almighty whom Jews generally refer to as Ha'Shem. I also tend to refer to him as *Our Father*, because he is our Maker and he leads us and guides us as a father. And I assume you know the Hebrew words that are translated in the King James Version of the Bible as LORD and Lord. Finally, I will also quote from the Stone Edition of the Tanakh.

Regarding your first question—where did I get the idea that the Messiah would suffer and die for our sins—there are (1) simple, (2) beautiful, and (3) complex answers.

The simple answer is that we have known from the Beginning that a *minchah* הָחְנִמ offering is no basis for forgiveness of sin (Genesis/Beresheit 4:3).

Genesis 4:3

And in process of time it came to pass, that Cain brought of the *fruit of the ground* an offering unto the LORD.

Genesis 4:5

But unto Cain and *to his offering He had not respect*. And Cain was very wroth, and his countenance fell.

We can summarize my point referring to the Brit Chadasha (which is entirely consistent in it's summary of Leviticus):

Hebrew 9:22

And almost all things are by the law purged with blood; and *without shedding of blood is no remission* (of sin).

There are other intuitive answers. David knew that the LORD would not suffer his soul to remain in the grave.

Psalms 49:15:

But *God will redeem my soul* from the power of the grave: for he shall receive me. Selah.

While David knew that the *Almighty himself* would redeem his soul, he also knew that men of their own means or good qualities/behavior cannot redeem anyone's soul…

Psalms 49:6

They that trust in their wealth, and boast themselves in the multitude of their riches;

Psalms 49:7

None [of them] can by any means redeem his brother, nor give to God a ransom for him:

Psalms 49:8

(For the redemption of their soul [is] precious, and it ceaseth for ever).

And righteous Job also knew his Redeemer would be God and not man:

Job 19:25

For I know [that] my redeemer liveth and [that] he shall stand at the latter [day] upon the earth:

The phrase “Latter Day” here is ןוֹרֲחאַ, *'acharown*, which is used in the TANAKH several times for the *Last Day*, or *Latter Day*, or *Day of the Lord.* Clearly, no mere human is going to redeem Job several thousand years after this proclamation. It must the Son of Man (the Messiah) as referred to in:

Psalms 80:17

Let thy hand be upon the man of thy right hand, upon the son of man [whom] thou madest strong for thyself.

Psalms 144:3

LORD, what [is] man, that thou takest knowledge of him! [or] the son of man, that thou makest account of him! [Note who is addressed here!]

Also, please note in regard to your question about the Messiah uniting the Tribes of Israel (as referred to as Ephraim in Ezekiel 37), scholars among the Hebrew Roots community around the world refer to the Revelation of the Apocalypse by the Apostle John (Yochanan) as the *Het Galut*, which most believe describes the ingathering of the exiles in the Day of the Lord.

The way that the Almighty communicates the role of the Messiah most effectively is through the Lives of the Fathers. The last message that he sent through the Hebrew prophets came through Malachi.

Malachi 4:4

Remember ye the law [Torah] of Moses my servant, which I commanded unto him in Horeb for all Israel, [with] the statutes and judgments.

Malachi 4:5

Behold, I will send you Elijah the prophet before the coming of the great and dreadful day of the LORD:

Malachi 4:6

And he shall turn the heart of the fathers to the children, and the heart of the children to their fathers, lest I come and smite the earth with a curse.

His last message directed his children to the Torah (the instructions of Moses). In particular, I believe Malachi 4:6 tells us to focus on understanding the shadow pictures portrayed in the lives of the Fathers (Abraham, Isaac, and Jacob).

Also, remember that in Isaiah, our Father said 6 times, in messages equivalent to this:

Isaiah 46:9-10 Remember the former things of old: for I [am] God, and [there is] none else; [I am] God, and [there is] none like me, *Declaring the end from the beginning, and from ancient times [the things] that are not [yet] done*, saying, My counsel shall stand, and I will do all my pleasure.

He told us (the children) to turn back to the Fathers.

And he also directed us (by implication in Malachi 4:5) to Elijah ha'Tishbi.

So I am speaking to you, here at the end of the age [approaching 6000 years of God's works upon the earth], and we are told to look back to the Torah for understanding. We are told to turn back to the Fathers for understanding. Who are the Fathers? The are Abraham, Isaac, and Jacob. And who is Elijah? Ha'Tishbi indicates that Elijah was not Hebrew, but was one of the *Whosoevers* who joined the Children of Israel through faith [Exodus 12:48, Exodus 12:49]. Here he represents the prophets, even as Moses represents the Torah.

So what do we learn about the work of the Messiah in the lives of the fathers and the prophets. I would say everything that can be learned without being there to observe the Messiah in his coming(s). For example, let us just identify a few of the works of the fathers that have prophetic implications.

Father Deed(s)	Scripture Reference
Abraham	
First Hebrew, crossed over (symbolically came out of Babylon)	Genesis 12:1, Genesis 12:4

Friend of God, did not withhold his only son when the Almighty asked him for a blood sacrifice

2 Chronicles 20:7, Galatians 3:6

And Abraham said, My son, God will provide himself a lamb for a burnt offering: so they went both of them together.

Genesis 22:8, Genesis 22:16

Isaac

Agreed to become a blood sacrifice at the request of the LORD

Genesis 22:8

Ya'akov

Endured Jacob's Trouble (~20 years service under broken promises but obtained wives and twelve sons).

Note that another Jacob's Trouble is yet to be fulfilled: Alas! for that day [is] great, so that none [is] like it: it [is] even the time of Jacob's trouble; but he shall be saved out of it.

Jeremiah 30:7

And God said unto him, Thy name [is] Jacob: thy name shall not be called any more Jacob, but Israel shall be thy name: and he called his name Israel. [Note that Jacob, like many believers after they find faith in the Messiah, acquired a new walk that was evident to all.]

Genesis 35:10

Redeemed the birth right from a profane man (Esau who not only despised the birthright but also could not be bothered to honor Abraham at his funeral)

Genesis 25:9

Moshe
The LORD thy God will raise up unto thee a Prophet from the midst of thee, of thy brethren, like unto me; unto him ye shall hearken.

Deuteronomy 18:15

Joseph put down into the pit for 3 days by his brothers, and put down into the pit (prison) again though false accusation, yet through Joseph all Israel is saved. But as for you, ye thought evil against me;

[but] God meant it unto good, to bring to pass, as [it is] this day, to save much people alive.

Genesis 50:20

According to the Torah portion, M'ketz (at the end), Joseph is the first born of Rachel, whose womb had to be opened by God. This would seem clearly to be a shadow picture of the Messiah Yeshua's birth.

Mathew 1:18 - 20

The slaying of the children in Bethlehem in an effort to kill the Messiah

Mathew 2:18

Shadrach, Meshach, Abednego, Sent into the fiery furnace (Sheol) and saved by the LORD himself (one like the Son of Man).

Daniel 3:14-:30

Daniel went down into the pit (Lions Den) and certain death

Daniel 6:16 -27

David In buying Mt. Moriah for 50 silver shekels instead of taking it from Araunah as a gift, David provided a title deed for the Temple Mount in the record of 2 Samuel 24:24, and took possession of the altar where Abraham was faithful to give his only son Isaac as a burnt offering on Mt. Moriah [this became the site of the Holy of Holies according to many scholars]. The almighty put his finger prints on this transaction via the meaning of the name, Moriah, which is "chosen by Jehovah."

2 Samuel 24:24, 2Chronicles 3:1

The Almighty gave David a vision of the Messiah's suffering.

Psalm 22

David knew he would be redeemed from the grave.

Psalm 49:15

Jonah Spent 3 days and 3 nights in the belly of the fish, and then was raised alive. The Messiah also spent 3 days and 3 nights in the grave.

Jonah 1:17-2:10[2]

---

2. Most Biblical students are unaware that Jonah was a relative of the Disciple Peter. Also, if one reads the diary of Christopher Columbus, it is clear that Columbus was a descendant of Simon Peter! Also, Jonah is pronounced Yonah in Hebrew! Yonah means Dove in Hebrew, by the way!

> Yeshua said no sign would be given to an evil and adulterous generation except the sign of the prophet Jonah who was 3 days in the belly of the whale. Christians keeping Easter cannot count to 3. Those who understand the moedim (Unleavened bread, Passover, first fruits) can easily find fulfillment of this prophecy in the scriptures.
>
> Matthew 12:39, Luke 11:29

Finally, the acid test for the Messiah's authenticity resides in the parameters Moshe established:

Deuteronomy 18:15 The LORD thy God will raise up unto thee a Prophet from the midst of thee, of thy brethren, *like unto me*; unto him ye shall hearken;

So the Messiah must exhibit traits and do exploits patterns after Moshe's life. Let's identify some specific traits and deeds to use as criteria for candidate Messianic figures. History records only two serious candidates for the Messiah: (1) Shimon Bar Kokhba, and (2) Yeshua of Nazareth.

Shimon Bar Kokhba led the revolution against the Romans (132-135 C.E.). There are few sources Legendary. Even his name is uncertain. His first name, Shimon, was found written on coins from the time of the revolt. His last name, however, is written with many variations, such as Ben Koziva or Bar Kozevah, in different documents. It is probable that his name was originally Bar Koseva, which is either his father's name or the name of a Judean settlement. It was likely changed to Bar-Kokhba during the revolt, as a reference to a verse in the Bible referring to the Messiah as a star (*kokhav*). This would be fitting as Bar-Kokhba was descended from the Davidic dynasty (which is the Messianic dynasty according to Jewish tradition) and the Messianic hopes of the nation centered around him. Bar-Kokhba was an imperious dictator who was in charge of both the army and the economy. During the Jewish revolt against Rome. He held the title of Nasi, which could be a Messianic allusion or could simply refer to the one in charge of army, administration and economy Bar-Kokhba relied on his own powers and, according to *aggada* (Rabbinic Tradition), when he went to battle *he asked God to "neither assist nor discourage u*s." From http://www.jewishvirtuallibrary.org.

My conclusion here is that a leader of Israel that declines prayer is a poor candidate for Messiah. Also, there exists a much more disparaging history of the Bar Kokhba revolt in the *The Annals of the* World by James Usshur. This work ascribes the difference between the Jewish total accumulated time calendar and total accumulated time as tabulated from historical records and extant calendars to rabbinical manipulation of Israel's historical records in order to bring Bar Kokhba's advent into concurrence with prophecies given by Daniel.

Yeshua of Nazareth does have traits that compare with Moshe. A detailed discussion of these differences follows.

Moshe	Yeshua
Saved from infanticide ordered by Pharaoh. Exodus 1:15 through Exodus 1:22 and Exodus 2:5	*Saved from the slaying of all children two years old and under in Bethlehem ordered by Herod* to annihilate the Messiah that the Magi journeyed to Judea to worship by a message from an angel in a dream. Mathew 2:13 and 16
Hosea 11:1: "When Israel was a child, I loved him, *and out of Egypt I called my son.* *Moses left Egypt as Pharoah drowned in the Red Sea* Exodus 15:4	Yeshua went down to Egypt and was there *until the death of Herod*: that it might be fulfilled which was spoken of the Lord by the prophet, saying, *Out of Egypt have I called my son*. Mathew 2:13 through Mathew 2:15

One of the most interesting correspondences between Moshe and Yeshua lies in the ethnicity of those who sought their lives as babes. We know that a Pharoah arose in the land that knew not Joseph, but where did he come from?

One of the most brilliant of the contemporary twentieth century sons of Judah is Immanuel Velikovsky. In his book, *Ages in Chaos*, Volume 1, 1952, he makes the case that the war-like Hyksos who migrated into Egypt just prior to the Exodus and rose into the power structure of Egypt (perhaps even to the person of the Pharaoh) were *Edomites*! This certainly explains the eagerness to enslave the Hebrews, murder Hebrew children, and endure judgments rather than surrender to Hebrews. It is also very clear that the Herod who sought Yeshua's life in murdering all the children in Bethlehem was an *Edomite*. It is also clear that this event did not take the Almighty by surprise. Mathew 2:18—In Rama was there a voice heard, lamentation, and weeping 2805, and 2532 great mourning, Rachel weeping [for] her children, and would not be comforted, because they are not.

Another declarative statement by Yeshua of his mission to re-unite Ephraim and Judah is recorded in John 10:16: And *other sheep I have, which are not of this fold*: them also I must bring, and they shall hear my voice; and *there shall be one fold*, [and] *one shepherd.*

This shall occur at the time of the fullness of nations (Melo ha'goyim–see Genesis 48:17-19). Our Father is planning for a crowd when this occurs:

Isaiah 54:2 Enlarge the place of thy tent, and let them stretch forth the curtains of thine habitations: spare not, lengthen thy cords, and strengthen thy stakes;

Isaiah 54:3 For thou shalt break forth on the right hand and on the left; and thy seed shall inherit the Gentiles, and make the desolate cities to be inhabited....[and through Isaiah 54:8].

Confirmation of the story of the Exodus

The Egyptian *Papyrus Ipumwer* provides a parallel confirmation and historical record of the events of the Biblical Exodus. The Shrine at El Arish also confirms the story of the Exodus.

Exodus	Papyrus Ipumwer
11:2—Let every man demand of his neighbor, jewels of silver, jewels of gold	3:23—Gold and jewels are fastened on the neck of female salves
12:33—And the Egyptians were urgent upon the people, that they might send them out of the land in haste; for they said we are all dead men	4.2—Forsooth, great and small say: I wish I might die.

13:21—And the Lord went before them by day in a pillar of a cloud to lead them the way; and by night in a pillar of fire to give them light…	7.1—Behold the fire has mounted up on high, to give them light; it's burning goes forth against the enemies of the land.
14:9—But the Egyptians pursued after them, all of the horses and chariots of Pharoah…and overtook them encamping by the sea, beside Pi-ha-hiroth (Khiroth). This is the same place that Pharaoh and his chariots met his fate in the Red Sea –the place of the whirlpool	The Papyrus records that the Pharoah was lost under unusual circumstances that never happened before. Papyrus 15:1 What has happened? Through it is to cause the Asiatics [Hyksos] to know the condition of the land.

There remains a black monolithic shine at *el-Arish* with *an inscription that Pharaoh Thom perished in a whirlpoo*l. A few lines later it says that he was thrown by the whirlpool high in the air. H departed to heaven; he was no longer alive. Note that Pi-Thom is the name of one of the two treasure cities built by the enslaved Hebrews; Pi-Thom means the abode of Thom. *El-Arish* in on the Sinai peninsula where it touches the Mediterranean Sea.

The Egyptian historian Manetho called the Pharoah who ruled just before the Hyksos invasion, Tutimaeus or Timaios.

Josephus, the Jewish historian, wrote that a people of ignoble origin from the east [e.g. the Hyksos], whose coming was unforeseen, had the audacity to invade the country, which they mastered by main force without any difficulty or even a battle.

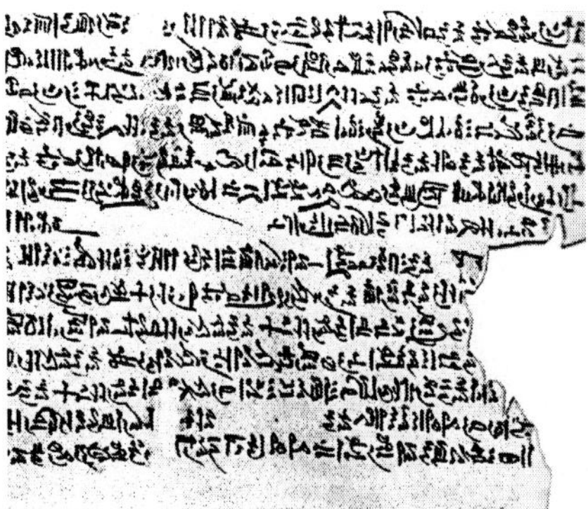

Portion of the Papyrus Ipumwer

Hard Proof of the Veracity of the Exodus Story

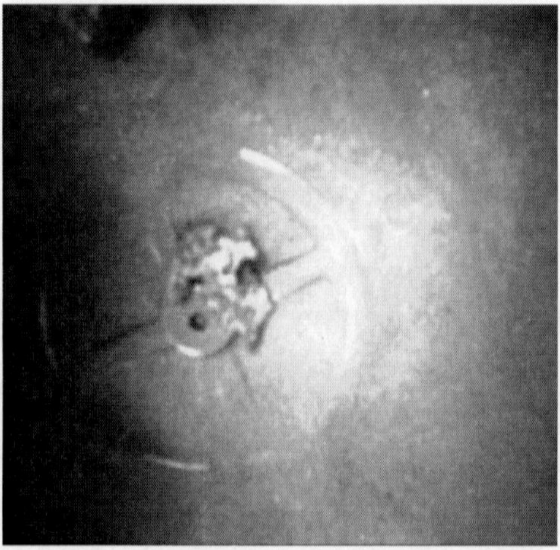

Archaeologist Ron Wyatt's photo of a four-spoke gold chariot wheel found on the bottom of the red sea on the underwater land bridge at Nuweiba (Ki-Ha-Hiroth). Only the Pharoah used a four-spoke wheel and only the Pharoah had gold wheels.

This demonstrates the veracity of the Exodus story in the Tanakh as well as the historical existence of horses and chariots and foundry work in Egypt at the time of the Exodus.

See The Exodus Case by Dr. Lennart Moeller of the Karolinski Institute, Stockholm, Sweden.

Understanding the Rise and Fall of the Hyksos

<u>A</u>

Chronology Table of Israel in Egypt				
Egypt	BC	Periods		Biblical Account
A people from Arabia bordering upon Egypt, called Hyksos by the Egyptians, invaded Egypt. They took Memphis and took over all of Lower Egypt bordering on the Mediterranean Sea.	2084	Saltis, their 1st king, reigned 19 years.	The Annals of the World, James Ussher, 1920c AM, 2630 JP (Manetheo 1:81 and Josephus , Apion 1.I.C14)	
	1910			Joseph sold into slavery
	1850			Jacob enters Egypt
			Genesis 36:31	Edomite King's list

Chronology Table of Israel in Egypt				
Egypt	BC	Periods		Biblical Account
			Deut. 2:12	Edom destroys Horites (Horims) and Hurrians (cave dwellers)
	170 years	Bela son of Beor		King in Edom; Genesis 36:32
		Jobab, son of Zerah in Bozrah (Gen 36:33)		Second King in Edom, Edomites push East
Hyksos invade		Husham		
		Hadad 1		Midian smitten
		Hadad II		
XVIIth Dynasty Ahmose I	1580		Exodus 1:8-11	Hyksos expelled from Egypt
	1503	Birth of Aaron		
Thutmosis III and Thut Queen	1501	Pharoah commands boy babies drowned		

Chronology Table of Israel in Egypt

Egypt	BC	Periods		Biblical Account
And he (Pharoah) said unto his people, Behold, the people of the children of Israel are more and mightier than we: 10 Come on, let us deal wisely with them; lest they multiply, and it come to pass, that, when there falleth out any war, they join also unto our enemies, and fight against us, and so get them up out of the land. 11 Therefore they did set over them taskmasters to afflict them with their burdens. And they built for Pharaoh treasure cities, Pithom and Raamses. 12 But the more they afflicted them, the more they multiplied and grew. And they were grieved because of the children of Israel.		Israel enslaved; Jewish sources hold that Israel was in slavery for 200 years. The Exodus took place 210 years after Jacob moved his family to Egypt.		Exodus 1:9-12

Chronology Table of Israel in Egypt				
Egypt	BC	Periods		Biblical Account
80 years before death of Amenhotep II	1500	Birth of Moses		(Moses means to be drawn from)
	1460	Moses flees to Midian		
Egyptian enslavement of Israel began inW	1420	Exodus of Israel form Egypt (Second year of Thutmosis III)		Moses is 80 years old and Aaron is 83. Thutmosis III reign began in 1501 BC per Breasted Chronology.
	1380	Death of Moses		
	1380	Joshua takes Jericho		Israel walks around Jericho 7 times and gives a mighty shout, and the walls come tumbling down
Amenhotep IV	1375			
	1374	Joshua		Most of Canaan taken by Israelites

Avaris was the Hyksos' capital city.
It's remains - Tell Al Dab'a –are shown above.

Hyksos and Edomites Characteristics in Common

1. Both are Semites (Semetic language and names)
2. Both have Hebrew characteristics
3. Both Have Hittite traits
4. Both appear to have been shepherds
5. Both are Arabians
6. Both lived easterly from Lower Egypt
7. These peoples seem to be called Hyksos by Egypt and Edom by Hebrew or Biblical Sources
8. Only EDOM fits the Hyksos traits

What would the Egyptians call this mixed horde breaking into and sweeping over the Delta region of Upper Egypt?

1. Arabian (Ishmaelite from Northern Arabia)
2. Asiatics that had Hebrew and Hittite blood. It is likely that Hittites from Canaan were part of the invaders.
3. Phoenecians of Mixed Cannanite and Hebrew stock
4. Rulers of Countries (they already ruled over their native lands).

This appellations are exactly the terms used by the Egyptians describing the Hyksos.

David J. Gibson postulates that the name of the Hyksos kings in the Egyptian record should parallel the names of the last Edomite kings in scripture.

Edomite King		Hyksos King	
Shaul	Could have turned to the Egyptian Delta after finish-ing conquests in the Euphrates region	Salatis	On the border nearest the delta Salatis founded his capital city of *Avaris* (Tanis or Zoan). Phoenetically similar to Avith or Avim or Acies in SW Palestine?)
Baal-hanan	Son of Achor Genesis 36:38	Beon	Reigned 44 years per Manetheo
Hadar or Hadad II	Hadar in Genesis Wb tahWut Hadad in 1 Chronicles	Apachnias	Wife was Queen Mehetebe 1

Other Hyksos Kings listed by Josephus (see *Josephus against Apion* I:13):

Apophia (1) who reigned sixty-one years; (2) Jonias (John or Khian) who reigned fifty years and one month [Joniaa ,or Khian, is one of the monuments found in widely scattered points from Southern Egypt to Crete and to Bagdhad]. (3) Aseis who reigned forty-nine years, two months

Thus far, there is no direct mention of horses in Egyptian or Hyksos records, albeit that the Hyksos are recorded as being horsemen and the bringer of chariots to Egypt. However, in the record of Jacob's funeral and return of his body to Israel, horses were in use. Bringing chariots means bringing the art of iron smelting and forging as well. Also, it should be considered that some of the Edomites settled in the region of Seir and Petra where much mining and smelting and forging of iron took place. The Horims also dwelt in Seir beforetime; but the children of Esau succeeded them, when they had destroyed them from before them, and dwelt in their stead.

And Joseph went up to bury his father: and with him went up all the servants of Pharaoh, the elders of his house, and all the elders of the land of Egypt,

Genesis 50:9

And there went up with him both chariots and horsemen: and it was a very great company.

Clearly, the horse and chariot had been brought into use when Joseph was still a great man in Egypt. This great man would have been quick to bring such treasures into use for the Royal household and other likely uses for beasts of burden.

Job (likely *Jobab*) also records that the use of horses for battle in his days:

Job 39:19

Hast thou given the horse strength? hast thou clothed his neck with thunder?

Job 39:21

He paweth in the valley, and rejoiceth in [his] strength: he goeth on to meet the armed men.

Job 39:22

He mocketh at fear, and is not affrighted; neither turneth he back from the sword.

Job 39:23

The quiver rattleth against him, the glittering spear and the shield.

Job 39:24

He swalloweth the ground with fierceness and rage: neither believeth he that [it is] the sound of the trumpet.

Job 39:25

He saith among the trumpets, Ha, ha; and he smelleth the battle afar off, the thunder of the captains, and the shouting.

Hath a Nation changed their gods?....Jeremiah 2:11

Hyksos Worship of False Gods

By the time we reach the 7th King in the order of Edomite Kings (starting at Genesis 36:31), Ba'al appears a god among the Edomites (the Hyksos). The 7th king took for a name –Baal-hanan (*Baal is gracious*). Job himself speaks of idolatry being practiced by some [Job 31:21-28].

Egyptian records indicate King Apophis made Sutekh his god.

It is also clear that the Almighty Yehovah was very displeased with Esau and his kingdom (Edom). In Psalm 83, The Tabenacles of Edom, the Ishmaelites, Amalek, and the Phillistines are grouped together as enemies of Israel and Almighty God. The Almighty says "Fill their faces with shame that they may seek thy name, O LORD."

It is also worthy of consideration that the Hyksos kings and the Edomites have certain deplorable characteristics in common. It may be presumed that the Hyksos persuaded the Pharoah Thutmosis to kill the Hebrew baby boys in 1501 BC.

Exodus 1:16

And he said, When ye do the office of a midwife to the Hebrew women, and see [them] upon the stools; if it [be] a son, then ye shall kill him: but if it [be] a daughter, then she shall live.

Exodus 1:22

And Pharaoh charged all his people, saying, Every son that is born ye shall cast into the river, and every daughter ye shall save alive.

There is one other case where an Edomite chose to kill babies to ensure the viability of his rule.

In Matthew 2:16, it is recorded that Herod the Idumean "sent forth and slew all the children that were in Bethlehem and it's suburbs from two years old and under." It is no wonder that the Almighty declared in Malachi 1:3, And I hated Esau and laid his mountains and his heritage waste for dragons of the wilderness. And it is recorded again in Romans 9:13, Esau have I hated.

Edom means the color red. Edom is derived from Adam (meaning red). Note that Esau killed Nimrod according to the Book of Jasher. And Rebekah told Jacob, "Behold, thy brother 251 Esau 6215, as touching thee, doth comfort himself, [purposing] to kill thee [Genesis s 27:42]. The Edomites were killers across the board.

What follows below is a depiction of the original founding of the nations after the Flood of Noah. Note that after the Flood, Shem, son of Noah, became the father of all the righteous nations [see Genesis 9:26-27; 10:22 and 10:31, 11:10, 11:11, and 1 Chronicles 1:17 and 1:24]. Furthermore, Shem lived a very long life and was known to be the righteous King of Salem (Melech Tzedek) to whom Abraham paid tithes [Hebrews 7:1-7:2].

Expanse of the Hyksos Empire

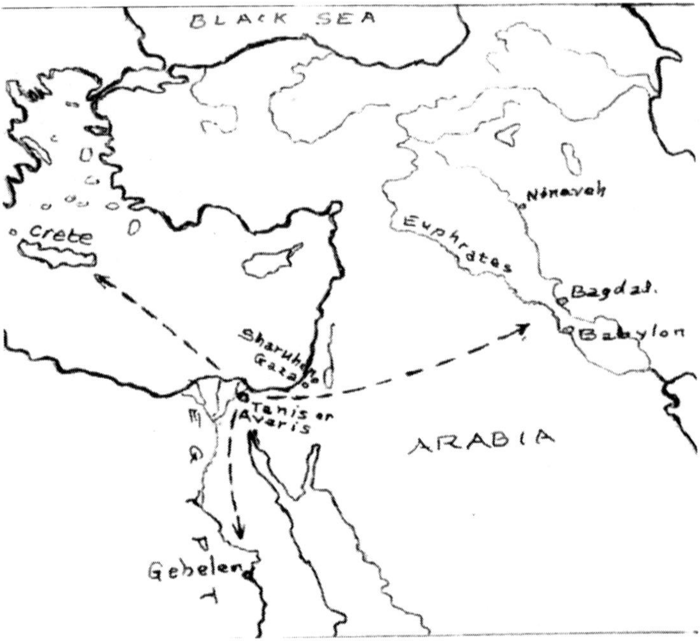

Total Elapsed Time and the Age of the Earth

I take this opportunity to provide two very credible sources for the total elapsed age of our universe (footnote 4) and to dispute the credentials of the Seder Olam Rabah. The Jewish *Seder Olam Rabah* (which is purported to be a total elapsed time calendar) is not consistent with Ussher's dating because years were subtracted from the *Seder Olam Rabah* as a result of the Rabbis' attempts to legitimize Shimon Bar Kokhba who led the rebellion against the Romans in 132-135C.E. There are few sources about Bar-Kokhba. Those that exist in Talmud, Midrash, and Church Father Euseblus' works are largely legendary.

Even Bar-Kokhba's name is uncertain. His first name, Shimon, was found written on coins from the time of the revolt. His last name, however, is written with many variations, such as Ben Koziva or Bar Kozevah, in different documents. It is probable that his name was originally Bar Koseva, which is either his father's name or the name of Judean settlement. His first name, Shimon, was found written on coins from the time of the revolt. His last name, however, is written with many variations, such as Ben Koziv or Bar Kozevah, in different documents.

It was likely changed to Bar-Kokhba during the revolt as a reference to verse in the Bible referring to the Messiah as a star (*kokhav*). This would be fitting as Bar Kokhba was descended from the Davidic dynasty (which is the Messianic dynasty according to Jewish tradition) and the Messianic hopes of the nation centered around him.

Bar-Kokhba was an imperious dictator who was in charge of both the arm and the economy during the Jewish revolt against Rome. He held the title of *Nasi*, which could be a Messianic allusion or could simply refer to the one in charge of army, administration, and economy. Bar-Kokhba relied on his own powers and, according to *aggada*

(Rabbinic tradition), when he went to battle *he asked God to "neither assist nor discourage us (*From http://www.jewishvirtuallibrary.org).

My conclusion here is that a leader of Israel that declines prayer is a poor candidate for the Messiah. Also, there exists a much darker history of deeds associated with the Bar Kokhba revolt in the *The Annals of the World* by James Ussher. This work ascribes the difference between the Jewish total accumulated time calendar and total accumulated time as tabulated from historical records and extant calendars to rabbinical manipulation of Israel's historical records in order to bring Bar Kokhba's advent into concurrence with prophecies given by Daniel. Specifically, Ussher attributes a 165 year discrepancy in the *Seder Olam Rabbah* —the book of the order of the world— apparently introduced by its author, Rabbi Yosef ben Halfta.

Shimon Schwab postulates that Rabbis intentionally "covered up a certain historic period and suppressed all records and other material pertaining thereto."

Schwab, along with other commentators, further suggests that the reason that God directed the sages of the second century AD to become involved in falsifying the data was to confuse anyone who might try to use the data to predict the time of the messiah's coming.

This was supposedly done to honor Daniel 12: —"shut up the words, and seal the book, even to the end." Others state that this was done to conceal the fact that the Daniel

9:25 prophecy pointed directly to the coming of Yeshua of Nazareth's ministry in 29 AD.

In addition to these historical documents, we have today scientific analyses that calculate the age of the earth in the neighborhood of six thousand years (supporting James Ussher's summation of years based on historical events). These calculations are based on the decay of primordial radionuclides in granite and determination that the amount of Carbon-14 and Helium is consistent across the fossil record. There should be virtually no Carbon-14 in fossils and diamonds given the C14 decay rate if the earth was billions of years old as postulated by proponents of evolution for the purposes of providing more time for evolution to result in new species and to draw attention away from their inappropriate application of age measurement methods such as carbon dating to materials older than the predictive capabilities of the dating method. Also, the relative absence of C14 in diamonds which shows conclusively that the time since the creation is about 6000 years. Finally, measurement of the daughter products from decay of radionuclides in granite mountains in New England proves unequivocally that the age of the earth is ~6000 years (not billions of years). I would highly recommend this work—*Thousands... Not Billions*—to those interested. It is produced by the Institute for Creation Research as a video, and is available in book form as *Thousands....Not Billions*: *Challenging the Icon of Evolution, Questioning the Age of the Earth*, by

Donald DeYoung. Also, the study, *Radiocarbon Dating—Its Limitations and Usefulness*, is very illuminating. Central Highlands Christian Publications, 289 Stag Rd, Kingston, Vic. 3364 Australia at http://www.chcpublications.net/radcarbn.htm.

The Name of the Messiah and God the Father

There is considerable misunderstanding related to the pronunciation of the Messiah's name and the name of God the Father. Regarding the pronunciation of the Messiah's name, commonly referred to as Jesus, it is necessary to go back to the process of translation of the Masoretic text into English by the King James translators.

The Hebrew text of the Old Testament is called the Masoretic Text because in its present form it is based upon the Masora—the Hebrew, textual tradition of the Jewish scholars known as the Masoretes (or Masorites). The Masorites were rabbis who made it their special work to correct the faults that had crept into the text of the Old Testament during the Babylonian captivity, and to prevent, for the future, its being corrupted by any alteration. They first separated the apocryphal from the canonical books, and divided the latter into twenty-two books, being the number of letters in the Hebrew alphabet. Then they divided each book into sections and verses.

There is a great difference of opinion as to when the Masoretic Text was written, but it was probably accomplished in the 10th–11th century. Several editions existed, varying considerably, but the received and authoritative text is that of Jacob ben-chayim ibn Adonijah, who carefully sifted and arranged the previous works on the subject. It was published in 1524. The King James Bible was an English translation of the Masoretic text, as were almost all Bibles, including the Geneva Bible that was published shortly before the King James and used by the pilgrims that came to America on the Mayflower.

In the first century, the Messiah's name was translated into Greek since the church at Constantinople was dominant. It was later translated into Latin as the Church of Rome got established. What was of importance here was maintenance of the phonetic pronunciation of his name. Most of us have never considered the difficulty of maintaining the original pronunciation of a name when transliterating the name to another language.

The Hebrew scriptures were first translated into Greek in what is called the Septuagint since it was produced by seventy Hebrew scholars at the request of Ptolemy II. The Septuagint has been very valuable in confirmation of the meaning of words and phrases of other Hebrew writings over the centuries. In the Septuagint, Joshua (as pronounced today, but in Hebrew it is spelled the same as the Jesus of Nazareth) is translated as IESOUS, in an effort to transliterate the sound of the original name.

The King James translators in an attempt to duplicate the original pronunciation of the name in English chose the J-sound, which was pronounced in the later 1500's and early 1600's with the Y-sound (Yuh) of today. So the name was spelled Jesus, but pronounced akin to IESOUS. Unfortunately, at the time of the publication of the King James Bible in 1611, the English consonants were undergoing an evolution in pronunciation. In this evolution, the J with a Y-sound evolved to the hard J-sound (Jay) of today and the intended IESOUS pronunciation ended up as the Jesus of today. Consequently, in an attempt to replicate the original pronunciation Hebraic roots believers of today pronounce the name as Ya'shua (meaning God's salvation in Hebrew). *Ya* is a diminutive form of YHVH (Yod He Vav is—the tetragrammaton which is not spoken aloud by the Jewish people because they hold it as too holy to be spoken aloud. And *Shua* means salvation. So, Jesus' real Hebrew name, Ya'Shua, means *God's Salvation*. This is clearly in concert with his role in life and as *our savior*. Note also that is clearly the appropriate name for the original Ya'Shua (Joshua of the Exodus), which was spelled the same way as Jesus-Ya'Shua in Hebrew). Today's widely used Jesus is simply a mispronunciation of the Savior's real name.

Finally, it is important to understand the correct pronunciation of the name of the God of our fathers. Devout Jews of today do not enunciate the name of their god based on the belief that the name of their god is too holy to speak aloud. As a result, most of the Hebrew people do not know the original pronunciation of the name of God. This pronunciation can be determined, however, from the Isaiah scroll found in the Qumram and now housed in a nitrogen atmosphere in the Israel Museum in Jerusalem. One of the Dead Sea Scroll translators of my acquaintance (who wishes to remain anonymous) has told me that a single vowel point in the Isaiah Scroll does reveal the correct pronunciation as Yehovah.

And one more point of interest for English-speakers. Most of us do not realize that our word, *alphabet*, is derived from the names for the first and last Hebrew letters –*aleph* and *bet*. Also, in reading the Hebrew text, the individual letters of the words enhance the meaning of the word. For example, the letter *Bet* means a house. And Aleph is the first letter of the Hebrew alphabet. The Rabbis believe the Aleph combines the divine, the spiritual and the physical world. In Revelation 22:13 Jesus refers to himself as the Aleph and Tav, the First and the Last.

Determining the Day of Jesus's (Ya'Shua's) Birth

While much of the world celebrates the birth of Jesus Christ on the 25 of December, this is actually the day when pagan gods (including Ra), can the actual day of Jesus' birth be determined from scripture? This question will be explored in some detail, and will yield a result that is quite intriguing. The first passage we will consider begins with the father of John the Baptist, Zacharias:

Luke 1:5—There was in the days of Herod, the king of Judaea, a certain priest named Zacharias, of the course of Abia: and his wife was of the daughters of Aaron, and her name was Elisabeth.

Luke 1:8–22—And it came to pass, that while he executed the priest's office before God in the order of his course...

Luke 1:23—And it came to pass, that, as soon as the days of his ministration were accomplished, he departed to his own house. Luke 1:24—And after those days his wife Elisabeth conceived...

The clue given to us here is that Zacharias was of the "course" of Abia (Abijah).

The 24 Courses of the Temple Priesthood

From Josephus, Antiquities of the Jews, Book 7, Chapter 14, Paragraph 7.

King David, being desirous of ordaining his son king of all the people, called together their rulers to Jerusalem, with the priests and the Levites and having first numbered the Levites, he found them to be thirty-eight thousand, from thirty-years old to fifty out of which he appointed twenty-three thousand to take care of the building of the temple, and out of the same, six thousand to be judges of the people and scribes, four thousand for porters to the house of God, and as many for singers, to sing to the instruments which David had prepared, as we have said already. He divided them also into courses: and when he had separated the priests from them, he found of these priests twenty-four courses, sixteen of the house of Eleazar, and eight of that of Ithamar; and he ordained that one course should minister to God eight days, from sabbath to sabbath. And thus were the courses distributed by lot, in the presence of David, and Zadok and Abiathar the high priests, and of all the rulers and that course which came up first was written down as the first, and accordingly the second, and so on to the twenty-fourth; and this partition hath remained to this day.

King David on God's instructions (1 Chr 28:11-13) divided the sons of Aaron into twenty-four groups (1 Chr 24:1-4) to setup a schedule by which the temple of the Lord could be staffed with priests all year round in an orderly manner. After the twenty-four groups of priests were established, lots were drawn to determine the sequence in which each group would serve in the Temple (1 Chronicles 24: 7-19). That sequence is as follows:

1 Chr 24:7	1. Jehoiarib	2. Jedaiah
1 Chr 24:8	3. Harim	4. Seorim
1 Chr 24:9	5. Malchijah	6. Mijamin
1 Chr 24:10	7. Hakkoz	8. Abijah

1 Chr 24:11	9. Jeshuah	10. Shecaniah
1 Chr 24:12	11. Eliashib	12. Jakim
1 Chr 24:13	13. Huppah	14. Jeshebeab
1 Chr 24:14	15. Bilgah	16. Immer
1 Chr 24:15	17. Hezir	18. Aphses
1 Chr 24:16	19. Pethahiah	20. Jehezekel
1 Chr 24:17	21. Jachim	22. Gamul
1 Chr 24:18	23. Delaiah	24. Maaziah

1 Chronicles 24:19—These were the orderings of them in their service to come into the house of the LORD, according to their manner, under Aaron their father, as the LORD God of Israel had commanded him.

Now each one of the twenty-four "courses" of priests would begin and end their service in the temple on the Sabbath, a tour of duty being for one week (2 Chronicles 23:8, 1 Chronicles 9:25). On three occasions during the year, all the men of Israel were required to travel to Jerusalem for festivals of the Lord, so on those occasions all the priests would be needed in the temple to accommodate the crowds. Those three festivals were Unleavened Bread, Pentecost, and Tabernacles (Deuteronomy 16:16).

The Yearly Cycle of Service in the Temple

The Jewish calendar begins in the spring, during the month of Nisan, so the first "course" of priests, would be that of the family of Jehoiarib, who would serve for one week, Sabbath to Sabbath. The second week would then be the responsibility of the family of Jedaiah. The third week would be the feast of Unleavened Bread (colloquially known as Passover), and all priests would be present for service. Then the schedule would resume with the third course of priests, the family of Harim. By this plan, when the twenty-fourth course was completed, the general cycle of courses would repeat. This schedule would cover fifty-one weeks or 357 days, enough for the lunar Jewish calendar (about 354 days). So, in a period of a year, each group of priests would serve in the Temple twice on their scheduled course, in addition to the three major festivals, for a total of about five weeks of duty.

The Conception of John the Baptist

Beginning with the first month, Nisan, in the spring (March-April), the schedule of the priest's courses would result with Zacharias serving during the tenth week of the year. This is because he was a member of the course of Abia (Abijah), the 8th course, and both the Feast of Unleavened Bread (15-21 Nisan) and Pentecost (6 Sivan) would have occurred before his scheduled duty. This places Zacharias' administration in the Temple as beginning on the second Sabbath of the third month, Sivan (May-June).

Zacharias was burning incense according to his Lot in the Temple, when an angel appeared to him. And the angel said unto him, "Fear not, Zacharias: for thy prayer is heard; and thy wife Elisabeth shall bear thee a son, and thou shalt call his name John."

Having completed his Temple service on the third Sabbath of Sivan, Zacharias returned home to his wife Elizabeth and soon conceived his son John [Luke 1:23-24].

Luke 1:23—And it came to pass, that, as soon as the days of his ministration were accomplished, he departed to his own house. Luke 1:24—And after those days his wife Elisabeth conceived ...

So *John (Yohanan) the Baptist* was probably conceived shortly after the third Sabbath of the month of Sivan.

The Conception of Jesus Christ

Now the reason that the information about John is important, is because according to Luke, Jesus was conceived by the Holy Spirit in the sixth month of Elisabeth's pregnancy:

Luke 1:24-27—And after those days his wife Elisabeth conceived, and hid herself five months, saying, thus hath the Lord dealt with me in the days wherein he looked on me, to take away my reproach among men. And in the sixth month the angel Gabriel was sent from God unto a city of Galilee, named Nazareth,to a virgin espoused to a man whose name was Yoseph, of the house of David; and the virgin's name was Mirium (not Mary)—Strong's H4813 and מִרְיָם in Hebrew.

Note that verse twenty-six above refers to the sixth month of Elisabeth's pregnancy, not Elul, the sixth month of the Hebrew calendar, and this is made plain by the context of the first chapter of Luke, verse 24 and again in verse 36:

Luke 1:36—And, behold, thy cousin Elisabeth, she hath also conceived a son in her old age: and this is the sixth month with her, who was called barren.

Mary stayed with Elizabeth for the last 3 months of her pregnancy, until the time that John (Yohanan -transliterated from Hebrew) was born.

Luke 1:56—And Mary abode with her about three months, and returned to her own house.

Luke 1:57—Now Elisabeth's full time came that she should be delivered; and she brought forth a son—John in English and Yohanan (phonetically) in Hebrew.

Now working from the information about John's conception late in the third month, Sivan, and advancing six months, we arrive late in the 9th month of Kislev (Nov-Dec) for the time frame for the *conception* of Ya'shua. It is notable here that the first day of the Jewish festival of Hanukkah, the Festival of Lights, is celebrated on the twenty-fifth day of Kislev, and Ya'shua is called the light of the world (John 8:12, 9:5, 12:46). This does not appear to be a mere coincidence. In the book of John, Hanukkah is called the feast of dedication (John 10:22). It is also the Festival of Lights. Hanukkah is an *eight* day festival of rejoicing, celebrating deliverance from enemies by the relighting of the menorah in the rededicated Temple, which according to the story of the Maccabees, stayed lit miraculously for eight days on only one day's supply of oil.

The Birth of John the Baptist

Based on a conception shortly after the third Sabbath of the month of Sivan, projecting forward an average term of about Ten lunar months (forty weeks), we arrive in the month of Nisan. It would appear that John the Baptist may have been born in the middle of the month, which would coincide with Passover and the Feast of Unleavened Bread. It is interesting to note, that even today, it is customary for the Jews to set out a special goblet of wine during the Passover Seder meal, in anticipation of the arrival of Elijah that week, which is based on the prophecy of Malachi:

Mal 4:5—Behold, I will send you Elijah the prophet before the coming of the great and dreadful day of the LORD:

Jesus identified John as the "Elijah" that the Jews had expected:

Mathew—17:10 And his disciples asked him, saying, Why then say the scribes that Elias [diminutive form of

Elijah] must first come? Mathew—17:11 And Ya'Shua answered and said unto them, Elias truly shall first come, and restore all things. Mathew—17:12 But I say unto you, That Elias is come already, and they knew him not, but have done unto him whatsoever they listed. Likewise shall also the Son of man suffer of them. Mathew—17:13 Then the disciples understood that he spake unto them of John [Yohanan] the Baptist.

The angel that appeared to Zacharias in the temple also indicated that Yohanan would be the expected "Elias":

Luke 1:17—And he shall go before him in the spirit and power of Elias, to turn the hearts of the fathers to the children, and the disobedient to the wisdom of the just; to make ready a people prepared for the Lord.

Finally, the Feast of Unleavened Bread begins on the fifteenth day of the first month, Nisan or Aviv, and this is the appropriate date for the birth of John the Baptist, the expected "Elijah."

The Birth of Jesus Christ

Since Jesus was conceived six months after John the Baptist, and we have established a likely date for John's birth, we need only move six months farther down the Jewish calendar to arrive at the date of the birth of Jesus. From the fifteenth day of the first month, Nisan, we go to the fifteenth day of the seventh month, Tishri. And what do we find on that date? It is the festival of Tabernacles! The fifteenth day of Tishri begins the third and last festival of the year to which all the men of Israel were to gather in Jerusalem for Temple services (Leviticus 23:34). It is also important to understand that the Christian Church has been seriously mislead regarding the event of the birth of Ya'Shua for centuries, maybe eons.

First of all, the only gods that were born on December 25 (before and after the progression of the equinox that was performed to get the calendar synchronized with the seasons after many centuries of use of the Julian calendar) were the pagan gods—Ra, Mithra, Tammuz, Ishtar, Nimrod, etc.

Secondly, Ya'Shua was born on the first day of the Feast of Tabernacles, as detailed below.

Immanuel

Isaiah 7:14—Therefore the Lord himself shall give you a sign; Behold, a virgin shall conceive, and bear a son, and shall call his name *Immanuel*.

Immanuel means "God with us". The Son of God had come to dwell with, or *tabernacle* on earth with his people.

John 1:14—And the Word was made flesh, and dwelt among us, (and we beheld his glory, the glory as of the only begotten of the Father), full of grace and truth.

The Feast of Tabernacles (Sukkot), occurs five days after the Day of Atonement, and is a festival of rejoicing and celebration of deliverance from slavery in Egypt (Leviticus 23:42-43). We are commanded to dwell in booths (Sukas) during this Moed. The *Modedim* [*appointed times* in Hebrew] of the Lord are spelled out in Leviticus 23 [the heart of the Torah]. The Modedim [appointed times of the Lord] of Leviticus 23 are still *mandatory* for all believers and are wonderful experiences.

The Feast of Tabernacles is a party for his people. We get to sleep under the stars. All believers are dwelling in Sukas (tents). *Strong drink* is specifically allowed…that is, you can convert the increase of your fields into money and use it buy strong drink for yourself and your household if that is what your soul desires [Deuteronomy 14:26]. And we are expected to give to the less fortunate and those without income like the Levites so that everyone has the means to enjoy the Feast.

And here is the time to correct the awful error of the Christian Church generated because of it's confusion of the ancient holidays of Saturnalia. Emperor Constantine, a pagan worshipper of Sol Invictus Mithra to his dying day, celebrated the rites of Saturnalia form the 17 of December through December 23, and what the Emperor did, everyone else was required to emulate. The festival of Saturnalia was purported to involve considerable debauchery. Under Constantine, the name of the holiday was changed to Christ-Mass. Constantine also ordered the death of his eldest sun Crispus and his wife Fausta. It is hard to see how this man could be a sincere Christian, especially given his edicts against the Jews at the Council of Nicea, and his forbidding of Keeping the Sabbath (a command of the Most High) and mandating the *Venerable Day of the Sun* to be kept instead.

Let us, for a moment, look at the reason for the error of the Church as it relates Christmas.

Luke 2:7—And she brought forth her firstborn son, and wrapped him in swaddling clothes, and laid him in a manger *because there was no room for them in the inn.*

Why was there no room at the inn? Mary and Joseph are always portrayed as near freezing, struggling against horizontal snow blowing in their faces, with Mary condemned to give birth in a freezing stable, all because Constantine wanted the Christmas to occur when the Romans were used to celebrating Saturnalia.

However, it is clear that Ya'Shua was born during tabernacles. Every Hebrew was commanded to dwell outdoors in a Suka (Lev. 23:42-43), not in a hotel. Joseph and Mary may have been without a Suka, but having to stay in a stable would not have been the freezing hardship the Church thinks it was.

Also of note is the fact that the Feast of Tabernacles is an *eight* day feast (Lev 23:36, 39). Why eight days? It may be because an infant was dedicated to God by performing circumcision on the *eighth* day after birth:

Luke 2:21—And when eight days were accomplished for the circumcising of the child, his name was called

Ya'Shua, which was so named of the angel before he was conceived in the womb.

Note: Modern medicine has determined that the blood-clotting capability of the human body achieves it's maximum on the eighth day after birth. This is the day that the Almighty commands us to circumcise our male children [Lev 12:3 In the eighth day the flesh of his foreskin shall be circumcised].

So the infant Ya'Shua would have been circumcised on the eighth and last day of the Feast of Tabernacles, a Sabbath day. This last day is considered a wonderful extension of the Feast of Tabernacles. It is called the Last great Day—Shemini Atzeret.

The Baptism of Jesus (Ya'Shua)

There is another indication in scripture as to when Jesus was born.

Mark 1:14-15—Now, after that John was put in prison, Jesus came into Galilee, preaching the gospel of the kingdom of God, and saying, The time is fulfilled, and the kingdom of God is at hand: Repent ye, and believe the gospel.

Jesus said this after his baptism, upon emerging from forty days in the wilderness, when he began his preaching ministry. The book of Daniel gives us the "time" or prophecy Jesus was speaking about.

Dan 9:25—Know therefore and understand that from the going forth of the commandment to restore and to build Jerusalem unto the Messiah, the Prince shall be seven weeks and threescore and two weeks: The street shall be built again, and the wall, even in troublous times.

Dan 9:26—And after threescore and two weeks shall Messiah be cut off, but not for himself: and the people of the prince that shall come shall destroy the city and the sanctuary; and the end thereof shall be with a flood, and unto the end of the war desolations are determined. Dan 9:27—And he shall confirm the covenant with many for one week: and in the midst of the week he shall cause the sacrifice and the oblation to cease...

It is very likely that by understanding this prophecy and date of the decree when it began, the wise men knew exactly when to look for the Christ child. The seventieth week of Daniel, a period of seven literal years, began with "Messiah the Prince." Messiah means anointed, and Jesus was publically anointed by the Holy Spirit at his baptism. Daniel 9:26-27 tells us that the Messiah would be "cut off" (crucified) in the "midst of the (seventieth) week", which is to say the Messiah would be crucified three-and-a-half years after his baptism.

Luke 3:22—And the Holy Ghost descended in a bodily shape like a dove upon him, and a voice came from heaven, which said, "Thou art my beloved Son, in thee I am well pleased." Luke 3:23—And Jesus himself began to be about thirty years of age...

Luke hints that at his baptism Jesus became about thirty, so it was likely that his birthday coincided, more or less, with his baptism (A Levitical priest began his service at the age of thirty, Numbers 4:3). So his baptism agrees with the time of Tabernacles, because three years and six months later at Passover, Jesus was crucified exactly and precisely as Daniel had prophesied, in the midst of the seventieth week. Knowing the year of his baptism from understanding Daniel, the wise men needed only to subtract thirty from it to know the year the Messiah would be born.

The Star of Bethlehem

"In the Old Testament (Daniel 9) the Saviour's advent was more clearly revealed. The magi learned with joy that his coming was near, and that the whole world was to be filled with a knowledge of the glory of the Lord. The prophet Daniel who was the chief of the Magi of Babylon, made provision for the gifts that the Magi brought many years before the star was seen. To understand the basis for recognition of this star, refer to http://www.biblicalastronomy.com/ (there is a nominal fee for membership) or the works of E.W. Bullinger. Note that there is extant astronomy software that is capable of showing the approach and departure of this star at the time of the Messiah's birth.

Conclusion

The scriptural evidence demonstrates conclusively that Ya'Shua (Jesus) was born on the fifteenth day of the month of Tishri, on the first day of the Feast of Tabernacles, which corresponds to the September—October timeframe of our present calendar!

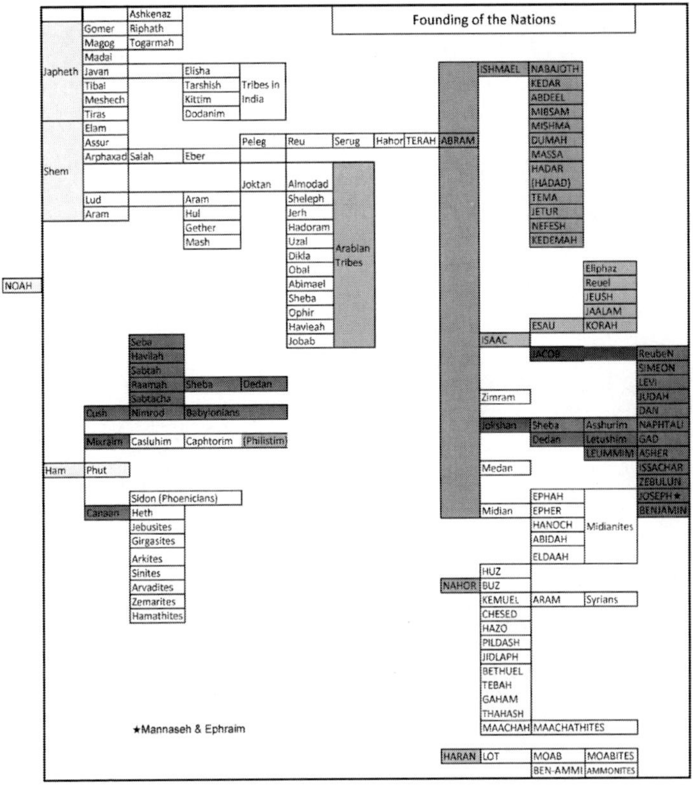

Final Words

A More Excellent Way

(a silly poem for Dan—Bless his memory—by his friend Rocky)

I suppose you heard when you were a kid
the strangest little rhyme 'bout what Moses did.
"Moses supposes his toeses are roses,
but Moses supposes erroneously."
It would seem like a "soap" if this prophet of hope
was so deluded he thought his feet exuded
the scent of flowers though he took no showers...
Yes, this holy man of mighty metacarpals—
who wandered forty years without many carpools—
was apparently confused about his own good news,
for the psalmist said, and it's been read and read,
"How beautiful upon the mountains
are the feet of those who bring good news"
—the real news is not inside shoes or even sandals:
it's that no matter what we lose on life's anvils,
there is a grand plan for a brand new Dan.
But thanks be to God that Moses did his duty
in spite of his quite odd belief in pedal beauty.
We too, live out and in our own illusions
and struggle daily with damnable confusions.

We often think we face too-tough trials
and wonder why we walk all uphill miles.
The problem's sometimes self-deceit
when we're convinced our feet smell sweet,
but for all of us who trudge like Moses
(under stress, in duress, lost in the wilderness,
without a bus, struggling through the Exodus)
God smooth's the path and daily shows us
a more excellent way.*

References

Josephus, *Antiquities of the Jews*, Book 7

Letter of Aristeas (No longer in print -only derivative copies available)

Thousands not Billions: Challenging the Icon of Evolution, Questioning the Age of the Earth, Donald DeYoung

Radiocarbon Dating: Its Limitations and Usefulness, CENTRAL HIGHLANDS CHRISTIAN PUBLICATIONS, 289 Stag Rd, Kingston, Vic. 3364 Australia at chcpublications.net/radcarbn.htm

Se*der Olam Rabbah* -the Book of the Order of the World, Rabbi Yosef ben Halfta

T*he Annals of the* World by James Ussher www.jewishvirtuallibrary.org

Discovery of a Large Early Edomite Empire, David J. Gibson, CanBook

The Exodus Case by Dr. Lennart Moeller of the Karolinski Institute, Stockholm, Sweden

http://splitrockresearch.org

The God of the Mountain, Penny Caldwell, Split Rock Research

Ages in Chaos, Immanuel Velikovsky, I: From the Exodus to King Akhnaton,

The Hyksos Invade Egypt.

Book of Jasher

Appendix

Abbreviated Curriculum Vitae, Mark A Robinson, Ph.D., Environmental Science

Education

- Arizona State University 1968, Iowa State University 1970, Wisconsin State University 1972, University of Houston 1975, California Coast University, BS, Engineering, 1982
- Colorado School of Mines, MS, Ecological Engineering, 1993
- Colorado School of Mines, Ph.D program, Environmental Science 1999-2000
- Lacrosse University, Ph.D Environmental Science, Summa Cum Laude, 2002

Vocations

- Radioactive and Hazardous Waste Investigator for the Department of Energy at the Rocky Flats Plant, the Waste Isolation Pilot Plant in Carlsbad, and the Los Alamos National Laboratory, New Mexico
- responsible for containers that intentionally or accidentally hold molten plutonium Los Alamos National Laboratory, 1987-2013

Avocation

Technical Expert Investigator for the International Atomic Energy Agency.

Investigated and resolved radioactive and hazardous waste problems in:

- Marcoule, France
- *Hashemite Kingdom of Jordan*
- Kenya
- Philippines
- Saudi Arabia

PUBLICATIONS

100 Grams of Uranium Equal 290 Tons of Coal, ISBN0-945318-00-6, M.A. Robinson, RandD Engineering, Kelso, WA, 1988

"Fraudulent, Substandard, and Mismarked Bolts", Nuclear News, M.A. Robinson, December 1987

"The Nuclear Fuel Cycle: Reprocessing vs. Direct Disposal", Colorado School of Mines, M.A. Robinson, May 1990

"The Toxicology of Naturally-occurring and Certain Important Man-made Radionuclides", M.A. Robinson, Colorado School of Mines, April 1991

"A Summary Comparison of the Environmental Effects of the Nuclear and Fossil Fuel Cycles in Electric Power Generation", Proceedings of the International Conference on Nuclear Power and the Environment, M.A. Robinson, Budapest, Hungary, October, 1992

"The Effect of Objective Criticality Safety Analyses on Radioactive Waste Management Activities", American Nuclear Society Annual Criticality Safety Meeting, April 1993

Technical Report Series No. 383, Characterization of Radioactive Waste and Waste Packages, M.A. Robinson, participating Author, International Atomic Energy Agency, Vienna, Austria, 1997

Safety and Performance Assessment, International Atomic Energy Agency, Manila, Philippines, February, 1995

Technical Reports Series No. 376, *Quality Assurance for Radioactive Waste Packages*, International Atomic Energy Agency, Vienna, Austria, M.A. Robinson, Primary Author, 1995

Radioactive Waste Disposal, International Atomic Energy Agency, M.A. Robinson, Author and Primary Lecturer, Manila, Philippines, February, 1995

Quality Assurance Principles in Radioactive Waste Management [Presentation Syllabus], M.A. Robinson, Author and Primary Lecturer, International Atomic Energy Agency, Manila, Philippines, February, 1995

Records and Documentation for Radioactive Waste Management and Disposal, M.A. Robinson, Primary Author, International Atomic Energy Agency, Manila, Philippines, February, 1995

IAEA-TECDOC-864, Requirements and Methods for Low and Intermediate Level Waste package Acceptability, International Atomic Energy Agency, Vienna, Austria, M.A. Robinson, Participating Author, February, 1996

IAEA-TECDOC-959, Inspection and Testing in Conditioning of Radioactive Waste, International Atomic Energy Agency, Vienna, Austria, Draft, Participating Author, August, 1997

LA-UR-97-3399, *Implementing Waste Minimization at an Active Plutonium Processing Facility: Successes and Progress at Technical Area 55 of the Los Alamos National Laboratory,* Los Alamos National Laboratory, J.J. Balkey, M.A. Robinson, MS, and J. Boak, 1997

LA-UR-03-2233, *Analysis of Plutonium Processing Systems as a Basis for Selecting Dispositions for Actinide-bearing Process Residues That Minimize Negative Quantifiable Consequences Including Proliferation risks, Radioactive Waste Generation, Personnel Exposure to Ionizing radiation, and Costs*, April 3, 2003Consequences

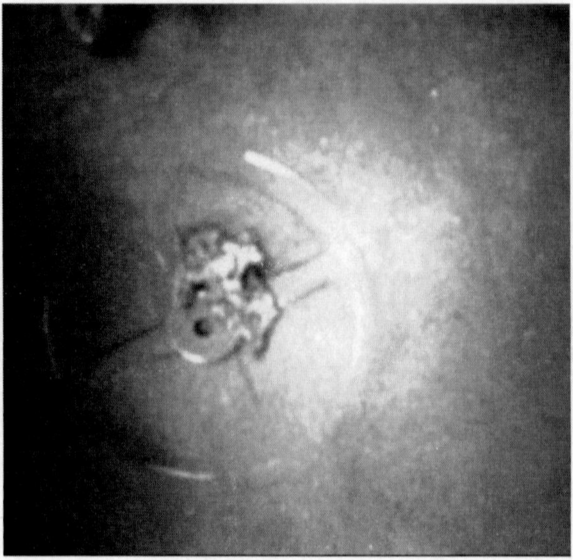

Golden Chariot Wheel (which belonged to Pharaoh as only Pharoahs had 4 spoke golden chariot wheels). Refer to The Exodus Case by Lennart Moeller of the karolinski Institute in Stockholm, Sweden (the institute which awards Nobel Prizes).

Mark Robinson, Ecological Engineer and Environmental Scientist, International Atomic Energy Agency, and Saudi Geologist from the Atomic Energy Research Institute (AERI) at Riyadh, at Bedouin Well -June 2009, 40 miles east of Mecca. Afternoon temperatures were about 128°F. The well was hand dug and ~90 feet deep. Water in the well is suitable only for camels. The area may receive less than an inch of rain in 100 years in contemporary centuries. It is possible that the Children of Israel might have visited this site in their wandering in the wilderness (Numbers 14:33). The children of Israel left chiseled images of their feet all over the wilderness from Mt. Sinai (Jabal Al Laws in NW Saudi Arabia not far from the Red Sea) to the Promised Land, making a record for the fulfillment of the promise of Yehovah in Joshua 1:33. Many of these imprints have been documented through the efforts of Jim & Penny Caldwell (splitrockresearch.org). At the top of the mountain range to the south of the photo location there is evidence of a dry lake with a dam constructed at the mouth of the canyon that drops into this plain. We were accompanied by a number of other technicians from AERI, making a party of 14 all together. When we returned to Riyadh, we had dinner in a traditional restaurant where all sit on the floor around a bowl of meat (lamb or goat) and rice that is eaten with fingers. After the meal I took a chance and gave the Saudi's a dissertation on the parable of the wheat & the tares given by Jesus in Matthew 13:24-30. The Saudi's were very excited to hear the words of Jesus who is mentioned in the Koran, but his words are not documented, and –in fact –reading his words is prohibited in Saudi Arabia.